Higher

Human Biology

2001 Exam

2002 Exam

2003 Exam

2004 Exam

2005 Exam

Leckie × Leckie

First exam published in 2001.
Published by Leckie & Leckie, 8 Whitehill Terrace, St. Andrews, Scotland KY16 8RN tel: 01334 475656 fax: 01334 477392
enquiries@leckieandleckie.co.uk www.leckieandleckie.co.uk

ISBN 1-84372-345-X

A CIP Catalogue record for this book is available from the British Library.

Printed in Scotland by Scotprint.

Leckie & Leckie is a division of Granada Learning Limited, part of ITV plc.

Acknowledgements

Leckie & Leckie is grateful to the copyright holders, as credited at the back of the book, for permission to use their material.
Every effort has been made to trace the copyright holders and to obtain their permission for the use of copyright material.
Leckie & Leckie will gladly receive information enabling them to rectify any error or omission in subsequent editions.

[BLANK PAGE]

FOR OFFICIAL USE

Total for Sections B & C

X009/301

NATIONAL QUALIFICATIONS 2001

MONDAY, 21 MAY 9.00 AM – 11.30 AM

HUMAN BIOLOGY HIGHER

Fill in these boxes and read what is printed below.

Full name of centre

Town

Forename(s)

Surname

Date of birth
Day Month Year

Scottish candidate number

Number of seat

SECTION A—Questions 1–30

Instructions for completion of Section A are given on page two.

SECTIONS B AND C

1 (a) All questions should be attempted.

(b) It should be noted that in **Section C** questions 1 and 2 each contain a choice.

2 The questions may be answered in any order but all answers are to be written in the spaces provided in this answer book, and must be written clearly and legibly in ink.

3 Additional space for answers and rough work will be found at the end of the book. If further space is required, supplementary sheets may be obtained from the invigilator and should be inserted inside the **front** cover of this book.

4 The numbers of questions must be clearly inserted with any answers written in the additional space.

5 Rough work, if any should be necessary, should be written in this book and then scored through when the fair copy has been written.

6 Before leaving the examination room you must give this book to the invigilator. If you do not, you may lose all the marks for this paper.

SCOTTISH QUALIFICATIONS AUTHORITY

SECTION A

Read carefully

1. Check that the answer sheet provided is for Human Biology Higher (Section A).
2. Fill in the details required on the answer sheet.
3. In this section a question is answered by indicating the choice A, B, C or D by a stroke made in **ink** in the appropriate place in the answer sheet—see the sample question below.
4. For each question there is only **one** correct answer.
5. Rough working, if required, should be done only on this question paper—or on the rough working sheet provided—**not** on the answer sheet.
6. At the end of the examination the answer sheet for Section A **must** be placed **inside** this answer book.

Sample Question

The digestive enzyme pepsin is most active in the

A mouth

B stomach

C duodenum

D pancreas.

The correct answer is **B**—stomach. A **heavy** vertical line should be drawn joining the two dots in the appropriate box in the column headed **B** as shown in the example on the answer sheet.

If, after you have recorded your answer, you decide that you have made an error and wish to make a change, you should cancel the original answer and put a vertical stroke in the box you now consider to be correct. Thus, if you want to change an answer D to an answer B, your answer sheet would look like this:

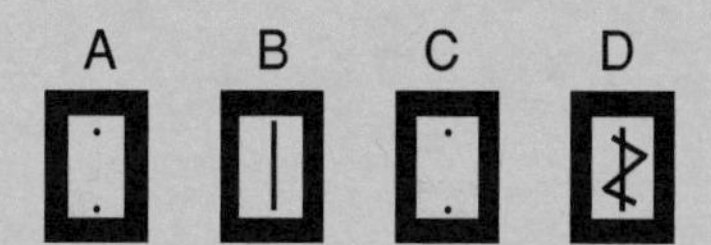

If you want to change back to an answer which has already been scored out, you should enter a tick (✓) to the **right** of the box of your choice, thus:

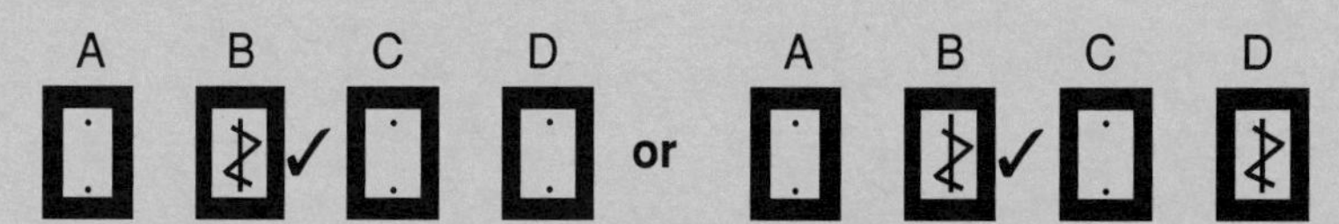

SECTION A

All questions in this section should be attempted.

Answers should be given on the separate answer sheet provided.

1. In respiration, the sequence of reactions resulting in the conversion of glucose to pyruvic acid is called

A the cytochrome system

B the TCA cycle

C the Krebs cycle

D glycolysis.

2. The diagram shows part of a liver cell with four parts labelled. In which part is most ATP produced?

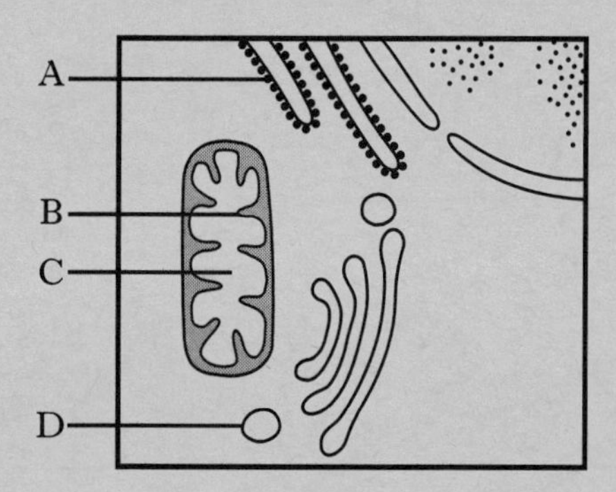

3. A DNA nucleotide could be formed from a molecule of phosphate together with

A ribose sugar and guanine

B ribose sugar and uracil

C deoxyribose sugar and guanine

D deoxyribose sugar and uracil.

4. If a DNA molecule contains 8000 nucleotides of which 20% are adenine, then the number of guanine nucleotides present is

A 1600

B 2000

C 2400

D 3200.

5. If the mass of DNA in a human liver cell is $6{\cdot}6 \times 10^{-12}$ g, the mass of DNA in a human sperm is likely to be

A $3{\cdot}3 \times 10^{-6}$ g

B $3{\cdot}3 \times 10^{-12}$ g

C $6{\cdot}6 \times 10^{-6}$ g

D $6{\cdot}6 \times 10^{-12}$ g.

6. A section of DNA has the following base sequence.

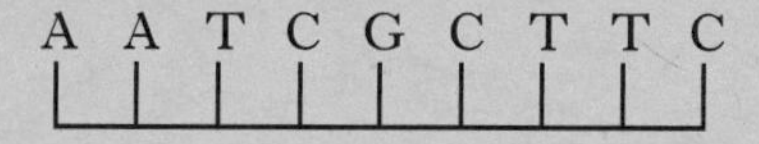

Identify the anti-codons of the three tRNA molecules which would align with the mRNA molecule transcribed from this section of DNA.

A AAU CGC UUC

B AAT CGC TTC

C TTA GCG AAG

D UUA GCG AAG

7. The cell organelle shown is magnified ten thousand times.

What is the actual size of the organelle?

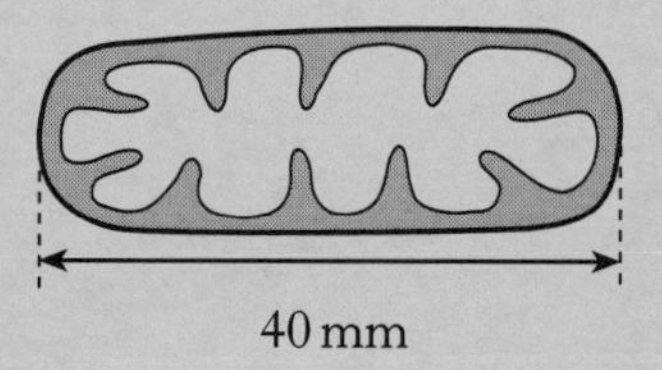

A $0{\cdot}04\,\mu m$

B $0{\cdot}4\,\mu m$

C $4\,\mu m$

D $40\,\mu m$

[Turn over

Questions 8 and 9 refer to the key shown below, used for the identification of carbohydrates.

1 { soluble 2
  { insoluble ....glycogen

2 { Benedict's test positive....................3
  { Benedict's test negative...................sucrose

3 { Barfoed's test positive.....................4
  { Barfoed's test negative ....lactose

4 { Clinistix test positive ....glucose
  { Clinistix test negative ....fructose

8. Which line in the table of results below is **not** in agreement with the information contained in the key?

	Carbo-hydrate	*Clinistix test*	*Barfoed's test*	*Benedict's test*
A	sucrose	not tested	not tested	negative
B	glucose	positive	negative	positive
C	fructose	negative	positive	positive
D	lactose	not tested	negative	positive

9. Maltose is a soluble carbohydrate which gives a positive result with Benedict's but not with Barfoed's reagent. With which carbohydrate in the key could maltose be confused?

A Fructose

B Glucose

C Sucrose

D Lactose

10. The stages of infection of a host cell by a virus are listed below.

1 Host cell bursts, releasing new viruses.
2 Host cell DNA is inactivated.
3 Virus binds to host cell and injects DNA.
4 Virus DNA directs synthesis of new viruses.

The sequence in which these events occurs is

A 3,2,4,1

B 1,2,4,3

C 3,4,2,1

D 2,4,3,1.

Questions 11 and 12 refer to the information below.

The diagram shows the chromosome complement of cells during the development of abnormal sperm.

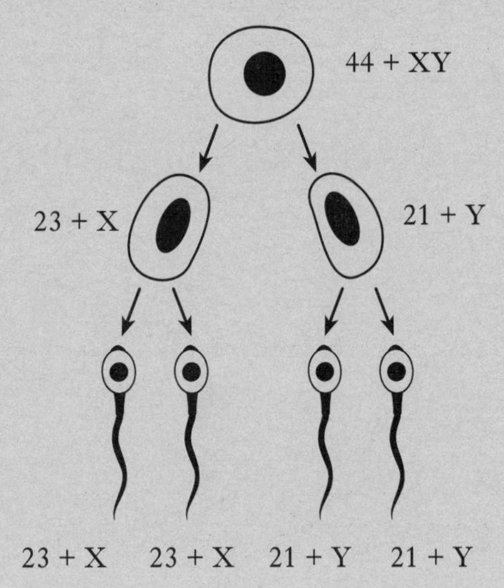

11. The diagram illustrates the effect of

A crossing over

B polygenic inheritance

C non-disjunction

D independent assortment of chromosomes.

12. A sperm with chromosome complement 23+X fertilises a normal haploid egg. What is the chromosome number and sex of the resulting zygote?

	Chromosome number	*Sex of zygote*
A	24	female
B	46	female
C	46	male
D	47	female

13. The colour of tooth enamel is a sex-linked characteristic. The allele for brown tooth enamel (e) is recessive to the allele for normal tooth enamel (E). The following family tree refers to this condition.

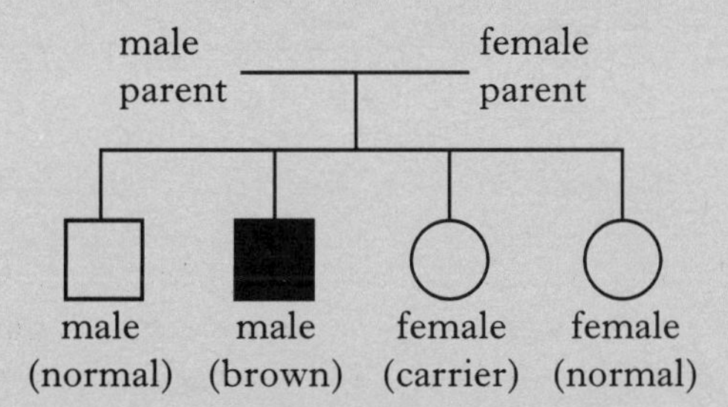

What are the genotypes of the parents?

A X^EY and X^EX^e

B X^EY and X^eX^e

C X^eY and X^EX^E

D X^eY and X^EX^e

Questions 14 and 15 refer to the following list of hormones.

A Follicle Stimulating Hormone (FSH)

B Luteinising Hormone (LH)

C Oestrogen

D Progesterone

14. Which hormone stimulates the production of testosterone by the testes?

15. Which hormone is produced by the corpus luteum?

16. Which of the following will **not** normally pass through the placenta between the mother and fetus?

A Oxygen

B Minerals

C Glucose

D Red blood cells

17. Fertility drugs may be used in the treatment of fertility to

A correct hormone imbalances

B reduce the pH of the oviduct

C stimulate mitosis in an egg

D protect sperm cells in the oviduct.

18. The durations of ventricular diastole and systole are shown below.

Diastole 0·4 seconds

Systole 0·2 seconds

What is the heart rate for this individual?

A 60 beats per minute

B 72 beats per minute

C 100 beats per minute

D 120 beats per minute

19. In which of the following pairs of tissues/organs, are red blood cells destroyed?

A Liver and lymph nodes

B Liver and spleen

C Bone marrow and duodenum

D Spleen and duodenum

20. Which of the following body fluids does **not** contain digestive enzymes?

A Saliva

B Gastric juice

C Pancreatic juice

D Bile

21. Which of the following results from an increase in the secretion of anti-diuretic hormone (ADH)?

A An increase in the permeability of the kidney tubules to water

B A decrease in the permeability of the kidney tubules to water

C An increase in the permeability of the glomerulus to water

D A decrease in the permeability of the glomerulus to water

22. Which of the following responses is caused by stimulation of the sympathetic nervous system?

A Increase in insulin production

B Increase in heart rate

C Increase in the flow of saliva

D Increase in peristalsis

23. A vertical section of the brain is shown in the diagram below.

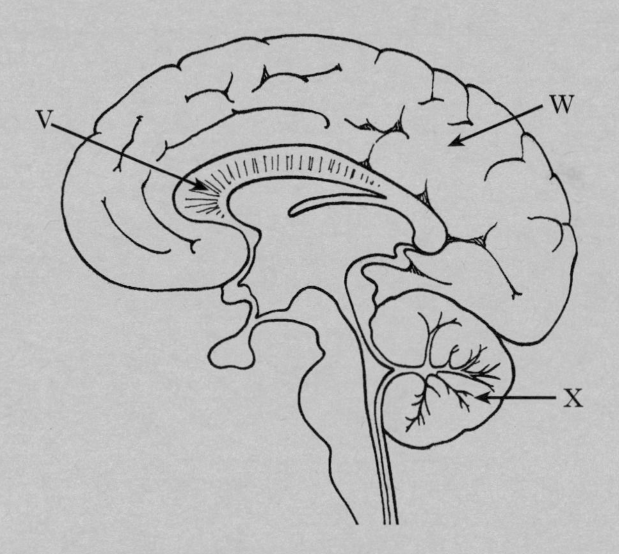

Which line of the table correctly labels the parts of the brain shown?

	V	*W*	*X*
A	corpus callosum	cerebellum	cerebrum
B	cerebellum	cerebrum	corpus callosum
C	corpus callosum	cerebrum	cerebellum
D	cerebrum	corpus callosum	cerebellum

24. Information is transferred between the two cerebral hemispheres by the

A corpus callosum

B medulla oblongata

C cerebellum

D hypothalamus.

25. A boy who is bitten by a large dog is subsequently frightened of all dogs. This behaviour pattern is an example of

A deindividuation

B extinction

C generalisation

D discrimination.

26. The development of phenotype is influenced by

A genetic factors only

B genetic factors and nutrition only

C the environment only

D the environment and genetic factors only.

27. The age structure of four different human populations is represented in the diagrams below. The bars indicate the relative numbers in each group.

Which diagram shows the population with greatest scope for growth?

A

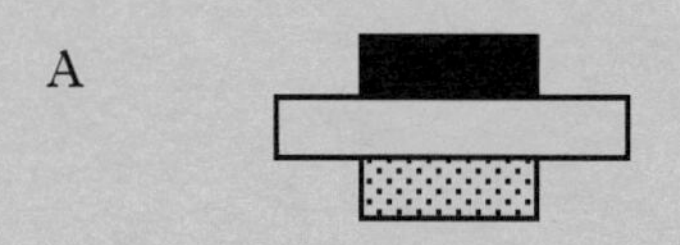

B

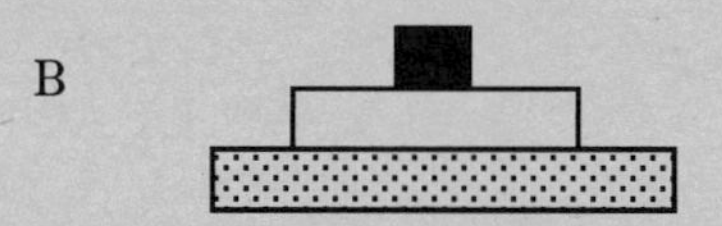

C

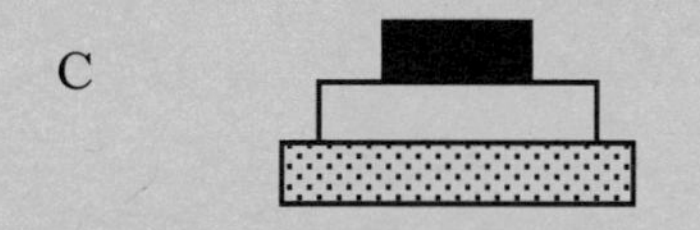

D

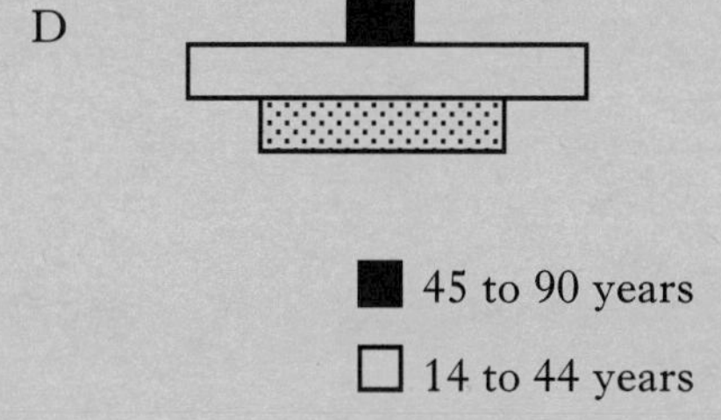

28. The interdependent biological and physical components in an area make up

A a habitat

B an ecosystem

C a food web

D a community.

29. What would be the effect of the discharge of raw sewage on the oxygen and nitrate concentrations of the water in a loch?

	Oxygen concentration	*Nitrate concentration*
A	increase	increase
B	increase	decrease
C	decrease	increase
D	decrease	decrease

30. The diagram shows a nitrogen cycle associated with the soil.

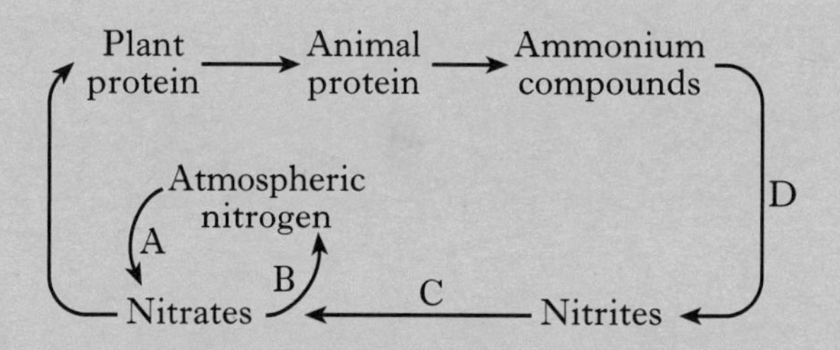

Which arrow indicates the activity of denitrifying bacteria?

Candidates are reminded that the answer sheet MUST be returned INSIDE this answer booklet.

[Turn over for Section B on *Page eight*

DO NOT WRITE IN THIS MARGIN

Marks

SECTION B

All questions in this section should be attempted.

1. The diagram below represents a reaction catalysed by an enzyme in the cytochrome system.

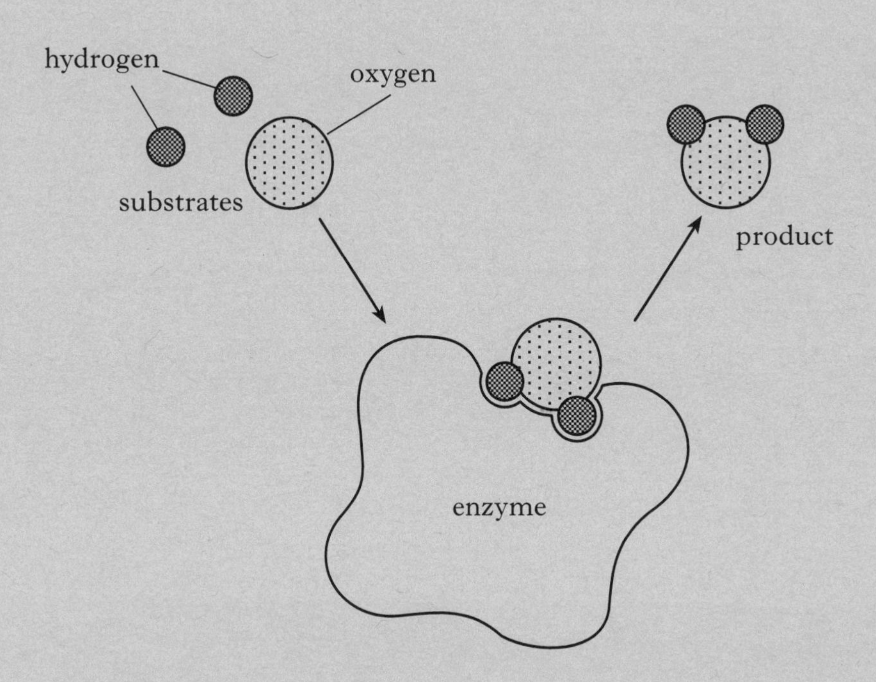

(*a*) (i) What name is given to the part of the enzyme where this reaction occurs?

________________________ **1**

(ii) In which organelle would this reaction take place?

________________________ **1**

(iii) Name the product of this reaction.

________________________ **1**

(*b*) Cyanide is a poison which inhibits this enzyme.

Suggest how cyanide is able to do this.

____

____ **1**

(*c*) Why do many enzyme-catalysed reactions require the presence of vitamins or minerals?

____ **1**

DO NOT WRITE IN THIS MARGIN

Marks

1. (continued)

(*d*) The graph shows the effect of increasing substrate concentration on the rate of this reaction.

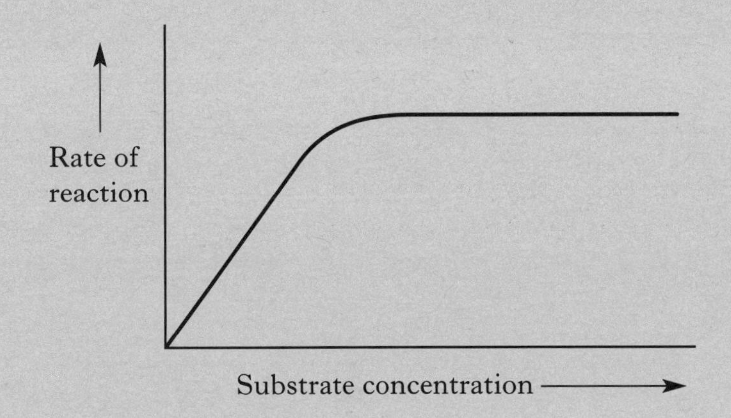

(i) Explain why the graph levels out at high substrate concentration.

____

______________________________ **1**

(ii) Assuming that the enzyme is operating at its optimum pH and temperature, suggest how the rate of reaction could be increased at high substrate concentrations.

____

______________________________ **1**

[Turn over

DO NOT WRITE IN THIS MARGIN

Marks

2. (*a*) The table below shows the relative concentrations of sodium and potassium ions in red blood cells and plasma.

	Sodium (units/litre)	*Potassium* (units/litre)
red blood cells	24	150
plasma	144	5

(i) Express, as simple ratios, the concentrations of sodium ions and potassium ions in the red blood cells and the plasma.

Space for calculation

Red blood cells : plasma

(1) sodium ____________ : ____________

(2) potassium____________ : ____________ **1**

(ii) Suggest how the red blood cells maintain the potassium concentration gradient.

____

____ **1**

(iii) When glucose is in short supply, the concentration of potassium in the red blood cells changes.

State whether the concentration will increase or decrease and give a reason for your answer.

Increase/decrease ____________

Reason ____________________________________

____

____ **2**

DO NOT WRITE IN THIS MARGIN

Marks

2. (continued)

(*b*) Three samples of red blood cells were placed in different concentrations of sodium chloride solution for two minutes. The results of this treatment, when viewed under the microscope, are shown in the diagram below.

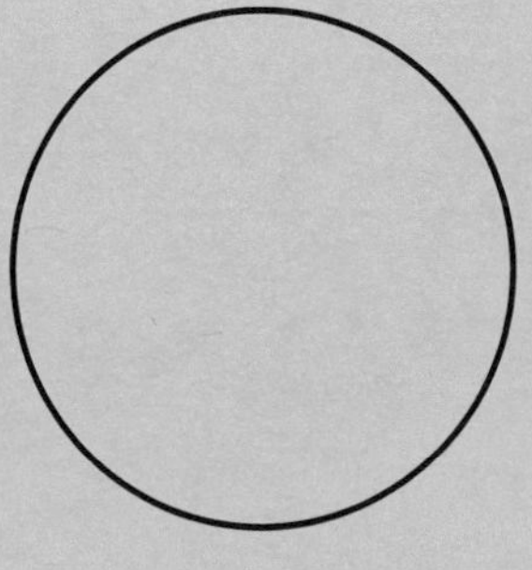

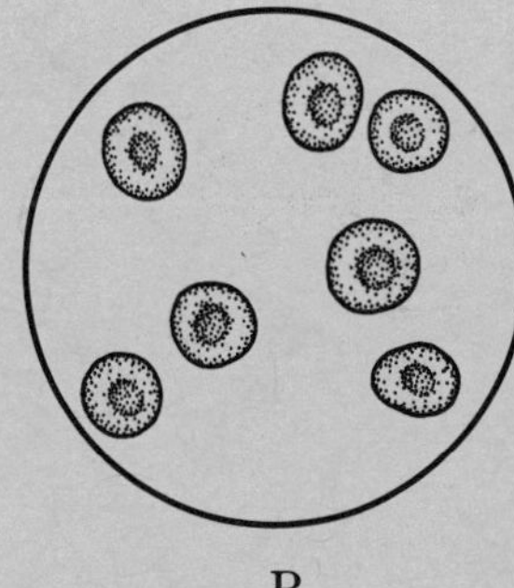

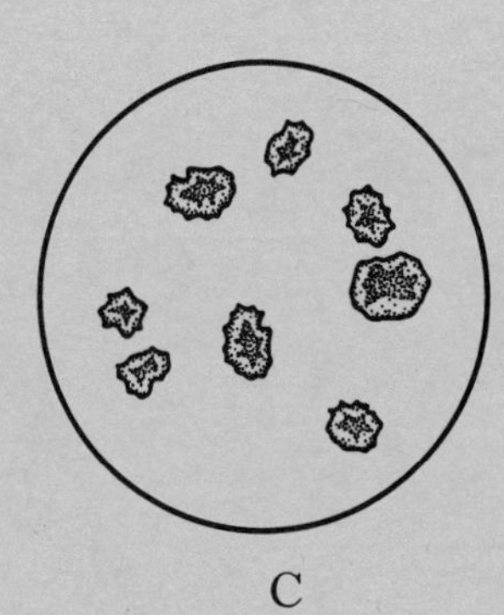

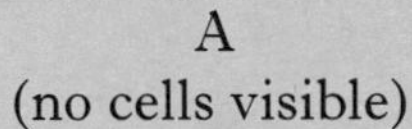

A (no cells visible) B C

Using the information above, explain the appearance of the cells in each diagram.

A ____

____

B ____

____

C ____

____ **3**

[Turn over

DO NOT WRITE IN THIS MARGIN

Marks

3. Antigens on the surface of red blood cells enable different blood groups to be identified.

Four types of blood group are A, B, AB and O.

The diagram shows antigens on a red blood cell and antibodies in the surrounding plasma.

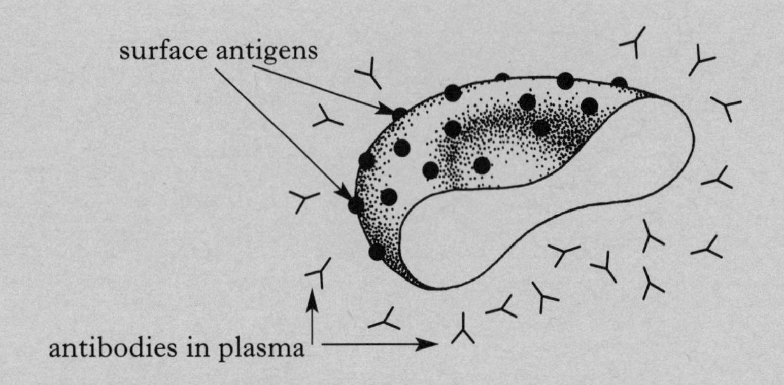

(*a*) Complete the table below to show the types of antigen and antibody present in individuals of each blood group.

Blood group	*Antigens present on surface of red blood cells*	*Antibodies present in plasma*
A	A	
B		anti-A
AB		
O		

2

(*b*) Which blood group(s) could be transfused safely into a person of blood group A?

______________________ 1

DO NOT WRITE IN THIS MARGIN

Marks

3. (continued)

(*c*) The gene for this blood group has three alleles. Alleles A and B are co-dominant to allele O.

A man, heterozygous for blood group A, and a woman of blood group AB, have children.

(i) State the genotypes of the parents.

female ____________________ male ____________________ **1**

(ii) Complete the Punnett square below to show the genotypes of their gametes and the genotypes of the children they may have.

female gametes / male gametes		

2

(iii) What is the percentage chance that a child of these parents would have blood group A?

____________________ % **1**

[Turn over

DO NOT WRITE IN THIS MARGIN

Marks

4. The diagram below illustrates the first stage of the process of amniocentesis. The fluid removed from the uterus contains fetal cells which can be grown, stained and examined.

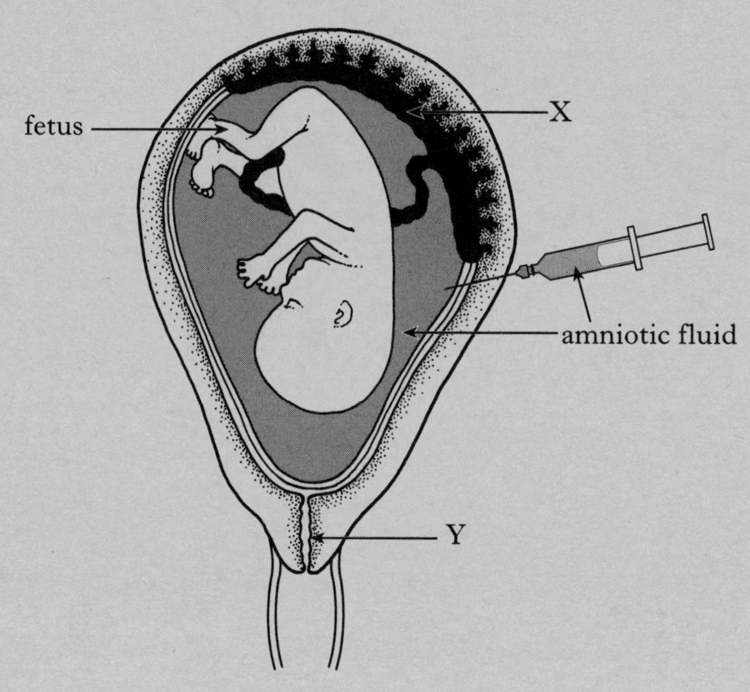

(*a*) (i) Identify structure X.

________________________ **1**

(ii) Name a hormone produced by structure X during pregnancy.

________________________ **1**

(*b*) (i) Identify structure Y.

________________________ **1**

(ii) Structure Y produces mucus. What change occurs to this mucus during the fertile period?

____ **1**

DO NOT WRITE IN THIS MARGIN

Marks

4. (continued)

(*c*) The diagram represents a photograph of stained chromosomes from a fetal cell.

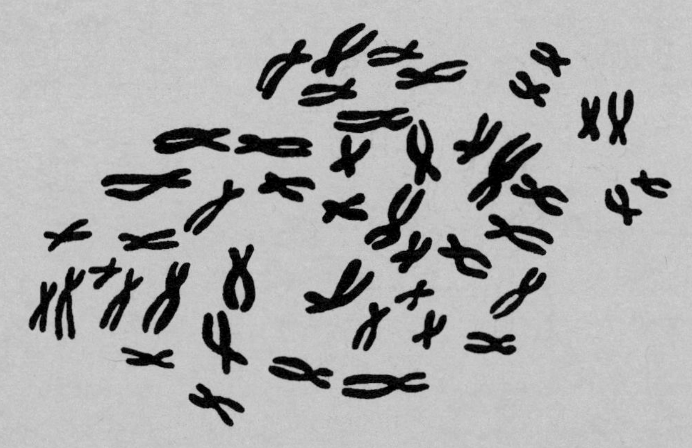

(i) The chromosomes may be cut out and arranged in homologous pairs.

Give **two** features of chromosomes which allow homologous pairs to be identified.

1 ____________________

2 ____________________ **1**

(ii) What name is given to a set of chromosomes arranged in pairs?

____________________ **1**

(iii) How could the sex of the fetus be identified from this paired arrangement?

____

____________________ **1**

(iv) What other information could be obtained which would be of value in pre-natal screening?

____

____________________ **1**

[Turn over

DO NOT WRITE IN THIS MARGIN

Marks

5. The diagram below shows the parts of a kidney nephron involved in ultrafiltration.

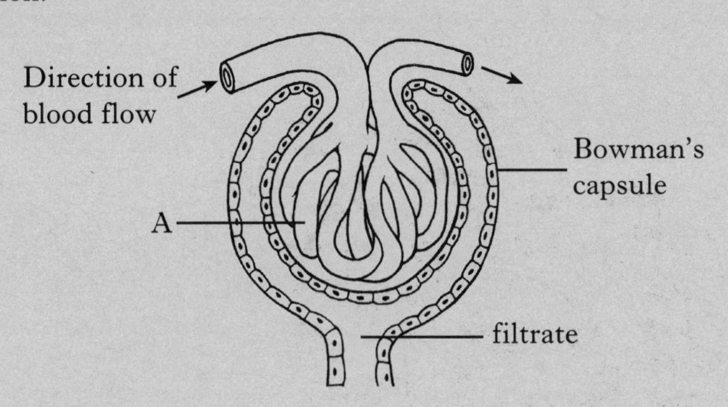

(*a*) (i) Part A consists of a bundle of capillaries. Name part A.

________________________ **1**

(ii) What feature, shown in the diagram, results in high blood pressure within structure A?

____ **1**

(iii) Into which part of the nephron does the filtrate flow immediately after leaving the Bowman's capsule?

________________________ **1**

(*b*) The table shows the composition of filtrate and urine.

Substance	*Mass in filtrate* (g/day)	*Mass in urine* (g/day)
Sodium ions	600	6
Potassium ions	35	2
Glucose	200	0
Urea	60	36
Water	180 000	1500

(i) Name the process which results in the differences between filtrate and urine, shown in the table.

________________________ **1**

(ii) What percentage of urea returns to the blood as the filtrate flows through the nephron?

______________ % **1**

(iii) Predict how the composition of urine would differ if the individual was an untreated diabetic.

____ **1**

DO NOT WRITE IN THIS MARGIN

Marks

6. The diagram below shows the parasympathetic nerve which runs between the central nervous system and the heart.

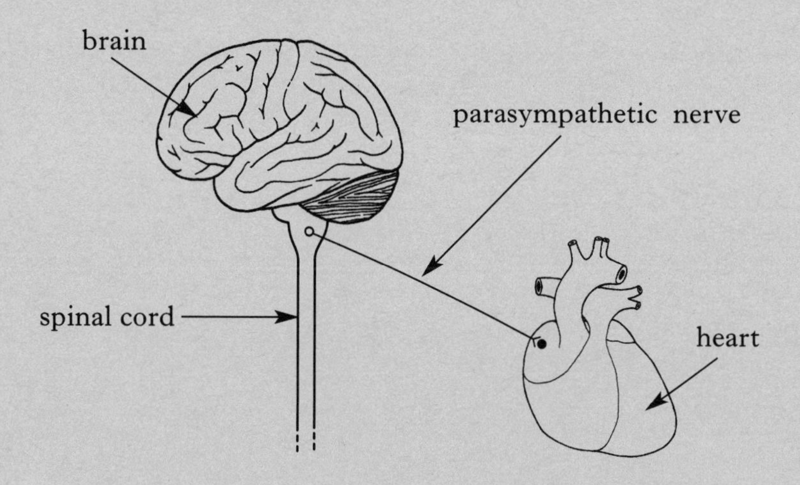

(*a*) (i) Which subdivision of the peripheral nervous system contains parasympathetic nerves?

________________________________ **1**

(ii) In which part of the brain does this parasympathetic nerve originate?

________________________________ **1**

(*b*) (i) Name the part of the right atrium which is stimulated by the parasympathetic nerve.

________________________________ **1**

(ii) State the effect of parasympathetic stimulation on the heart.

____ **1**

(*c*) Describe another effect which the parasympathetic nervous system has on the body.

____

____ **1**

[Turn over

7. In humans, alterations in the level of exercise bring about changes in pulse rate, stroke volume and ventilation rate. The level of exercise is measured as rate of oxygen uptake.

Graph 1 gives information about the heart. It shows how pulse rate and stroke volume change with the level of exercise in an individual.

Stroke volume is the volume of blood pumped from the heart in one beat.

Graph 1

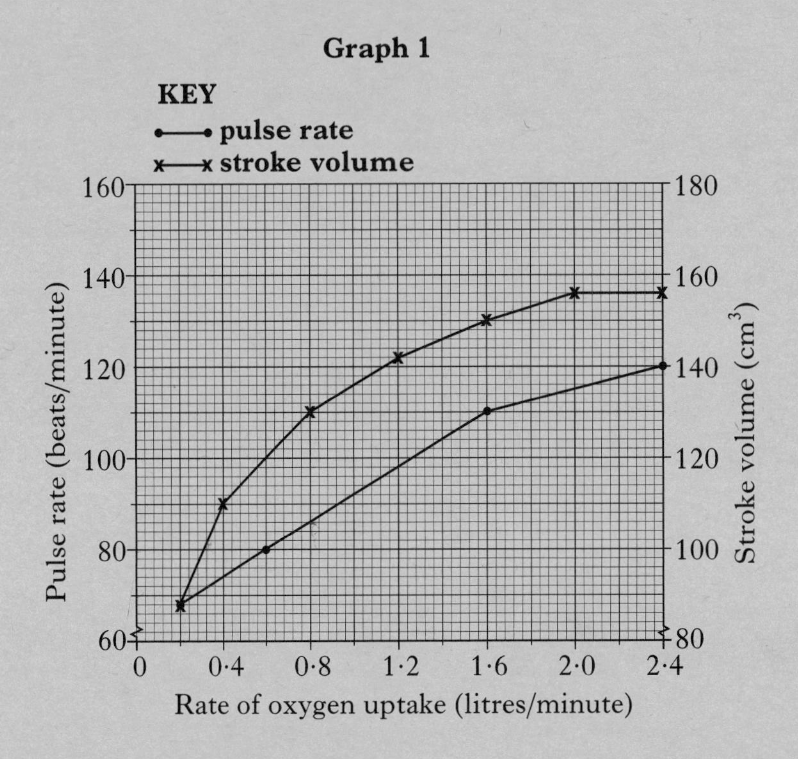

Graph 2 gives information about the lungs. It shows how the ventilation rate changes with the level of exercise in the same individual.

Ventilation rate is the volume of air inhaled during one minute.

Graph 2

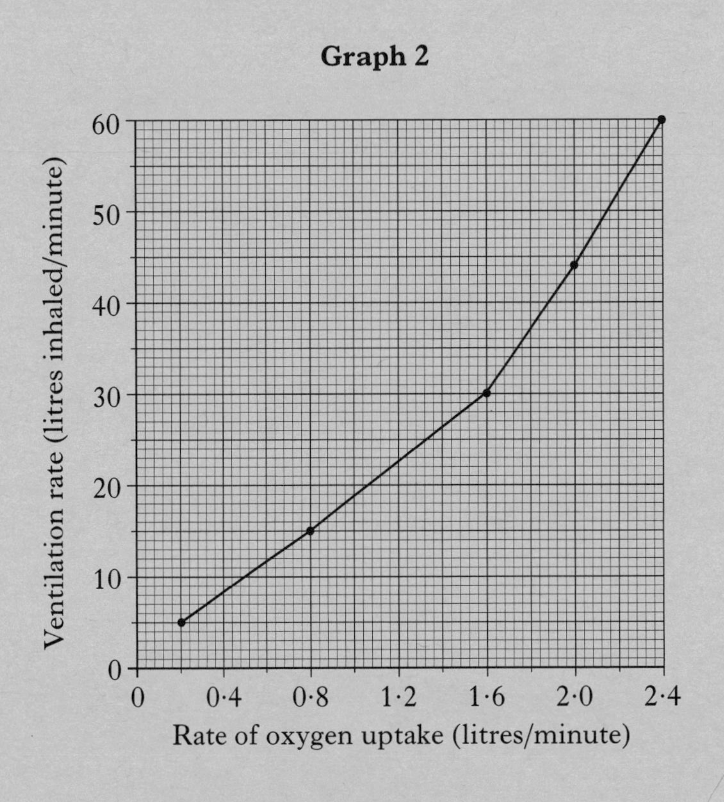

DO NOT WRITE IN THIS MARGIN

Marks

7. (continued)

(*a*) (i) With reference to **Graph 1**, what is the pulse rate and the stroke volume when the rate of oxygen uptake is 0·8 litres/minute?

Pulse rate ________________ Stroke volume ________________ **2**

(ii) What is the stroke volume when the pulse rate is 74 beats per minute?

____________ cm^3 **1**

(iii) What is the total volume of blood leaving the heart in one minute when the rate of oxygen uptake is 1·6 litres/minute?

Space for calculation

________________ litres/minute **1**

(iv) From **Graph 1**, compare the pattern of changes in pulse rate and stroke volume as oxygen uptake increases.

____

____

____ **2**

(*b*) (i) Fresh air contains 20% oxygen. From **Graph 2**, what is the volume of oxygen inhaled per minute when the rate of oxygen uptake is 1·6 litres/minute?

Space for calculation

________________ litres **1**

(ii) What additional information would be required to calculate the average volume of air taken in during each breath at any time?

____ **1**

(*c*) (i) With reference to **both graphs**, state the ventilation rate when this individual's pulse rate is 100 beats per minute.

____________ litres/minute **1**

(ii) Complete the table below by ticking the correct statement(s).

Statement	*Tick* (✓)
The rate at which pulse rate changes is highest at low rates of oxygen uptake.	
When ventilation rate doubles, the rate of oxygen uptake doubles.	

1

(*d*) Name the blood vessel which carries deoxygenated blood from the heart to the lungs.

________________________ **1**

DO NOT WRITE IN THIS MARGIN

Marks

8. Part of a neurone is shown in the diagram below.

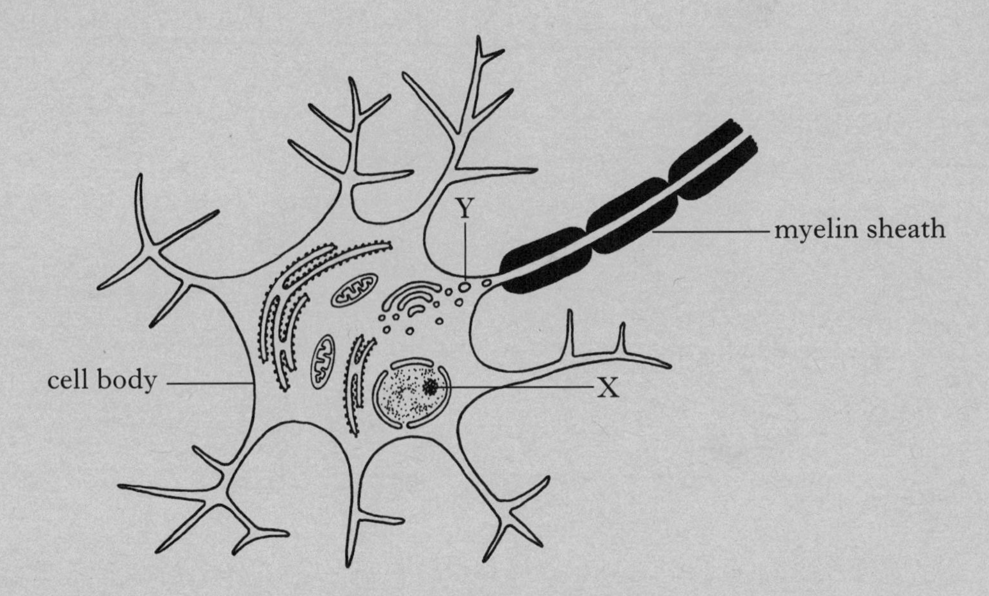

(*a*) State whether the neurone shown is a sensory or motor neurone and give a reason for your answer.

Type of neurone ______________________________

Reason for answer ______________________________

______________________________ **1**

(*b*) Name structure X and state its function.

Name ______________________________

Function ______________________________ **2**

(*c*) (i) Name structure Y.

______________________________ **1**

(ii) Similar structures are found in the synaptic knob. What do they contain?

______________________________ **1**

DO NOT WRITE IN THIS MARGIN

Marks

8. (continued)

(*d*) In the disorder Multiple Sclerosis, the myelin sheath is damaged by the body's own defence system.

(i) What effect does this have on the function of the nerve fibre?

____

____ **1**

(ii) What term is used to describe a disorder where the body's defence system destroys its own cells?

______________________ **1**

(*e*) Draw an arrow on the diagram to show the direction of an impulse in a dendrite. **1**

(*f*) Diverging neural pathways always contain the type of neurone shown opposite. Explain how diverging pathways allow humans to perform a task such as threading a needle.

____

____

____

____ **2**

[Turn over

DO NOT WRITE IN THIS MARGIN

Marks

9. "Fast foods" are now very much part of the culture of the developed world.

The table below gives information about a beef burger.

Beef burger

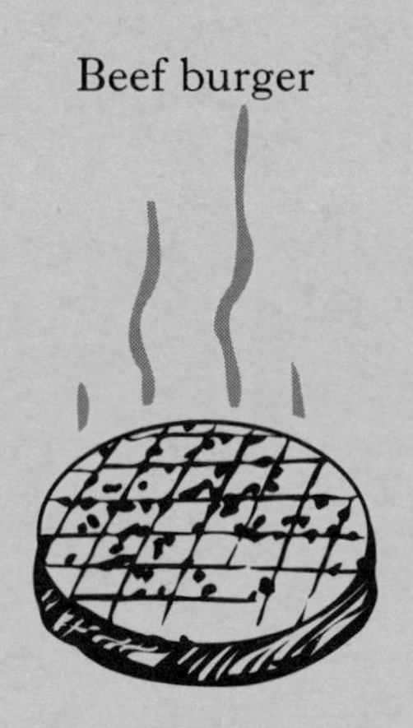

Nutritional analysis/100 g	
Energy	1500 kJ
Protein	12 g
Carbohydrate	8 g
Fat	30 g
Fibre	1 g
Sodium	1 g

(*a*) A boy ate little else but beef burgers every day.

With reference to the table, explain why the boy might suffer from malnutrition but not starvation.

____

____

______________________________ **2**

(*b*) Increased demand for cheap beef has had an impact on the natural ecosystems of developing countries.

Suggest how this demand affects natural ecosystems and local water supplies.

Natural ecosystems ______________________________

____

Local water supplies ______________________________

______________________________ **2**

(*c*) Why is the production of beef an inefficient use of land in a developing country where there is a large population to feed?

____

______________________________ **1**

DO NOT WRITE IN THIS MARGIN

Marks

9. (continued)

(*d*) The carbohydrate in the burger comes from wheat.

Modern varieties of wheat have been produced by selective breeding.

Describe an improvement brought about by selective breeding of crop plants such as wheat.

____

____

____ **1**

(*e*) (i) Pesticides are frequently applied to growing crops. What is a pesticide?

____ **1**

(ii) Describe **one** advantage of using pesticides.

____

____

____ **1**

(*f*) Genetic manipulation is now used to produce new varieties of organisms.

Describe **one** advantage of this technique compared to selective breeding.

____

____

____ **1**

[Turn over

DO NOT WRITE IN THIS MARGIN

Marks

10. The apparatus shown in **Figure 1** was used to investigate the effect of nitrates on the growth of grass. Grass seedlings were grown in seven different culture solutions. The experiment was repeated ten times.

Figure 2 shows the nitrate concentrations of the culture solutions and the results of the experiment.

Figure 1

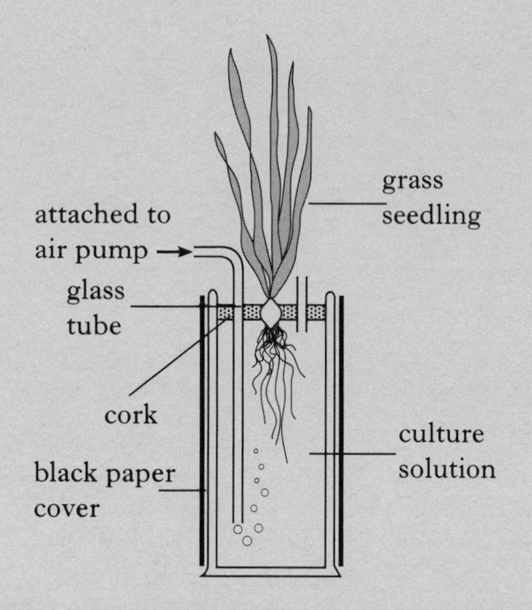

Figure 2

Culture solution	*Nitrate content* (g/litre)	*Average height of plants after 6 weeks of growth* (cm)
A	0	3
B	0·5	12
C	1·0	17
D	1·5	23
E	2·0	25
F	2·5	24
G	3·0	25

(*a*) Plot a line graph to illustrate the experimental results.

(Additional graph paper, if required, will be found on page 28.)

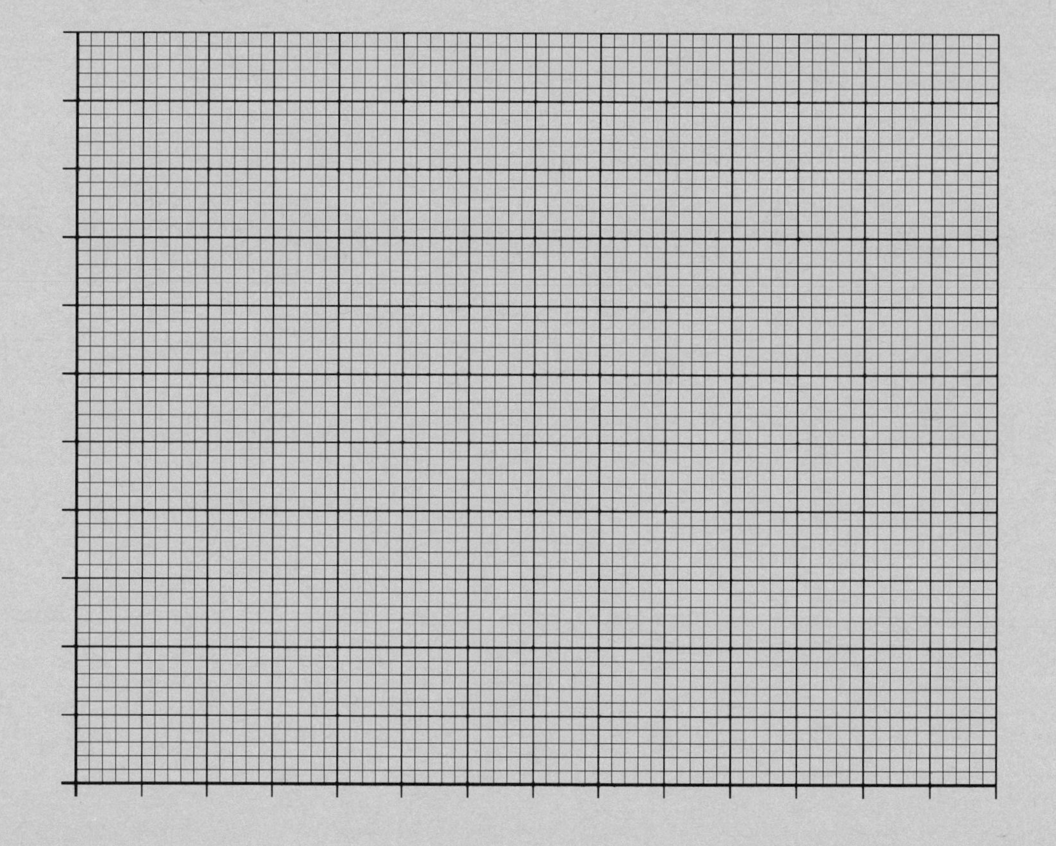

3

DO NOT WRITE IN THIS MARGIN

10. (continued)

Marks

(*b*) Name **two** variables which should be controlled in this experiment.

1 ____________________

2 ____________________ **1**

(*c*) During the experiment air was pumped into the jars through the glass tubes.

Suggest why this was necessary.

____

____________________ **1**

(*d*) What feature of the experiment makes the results more reliable?

____

____________________ **1**

(*e*) State **one** other feature of grass plants which could be observed or measured to assess the effects of nitrate.

____________________ **1**

(*f*) Predict how the results of this experiment would be different in culture solutions A, B and C if clover or pea plants had been used instead of grass plants. Give a reason for your answer.

Prediction ____________________ **1**

Reason ____________________ **1**

(*g*) A farmer wishes to purchase nitrate fertiliser. In what way could the information from this experiment be useful to the farmer?

____

____________________ **1**

[Section C begins on *Page twenty-six*

DO NOT WRITE IN THIS MARGIN

SECTION C

Marks

Both questions in this section should be attempted.

Note that each question contains a choice.

Questions 1 and 2 should be attempted on the blank pages which follow.

Supplementary sheets, if required, may be obtained from the invigilator.

Labelled diagrams may be used where appropriate.

1. Answer **either** A **or** B.

A. Give an account of memory under the following headings:

(i) encoding; **6**

(ii) storage; **2**

(iii) retrieval. **2**

(10)

OR

B. Give an account of immunisation under the following headings:

(i) artificial active immunity; **6**

(ii) artificial passive immunity; **2**

(iii) the impact of vaccination on childhood diseases. **2**

(10)

In question 2 ONE mark is available for coherence and ONE mark is available for relevance.

2. Answer **either** A **or** B.

A. Describe the biological basis of contraception. **(10)**

OR

B. Outline the involuntary mechanisms involved in temperature control. **(10)**

[*END OF QUESTION PAPER*]

DO NOT WRITE IN THIS MARGIN

SPACE FOR ANSWERS

Marks

DO NOT WRITE IN THIS MARGIN

Marks

SPACE FOR ANSWERS

ADDITIONAL GRAPH PAPER FOR QUESTION 10(*a*)

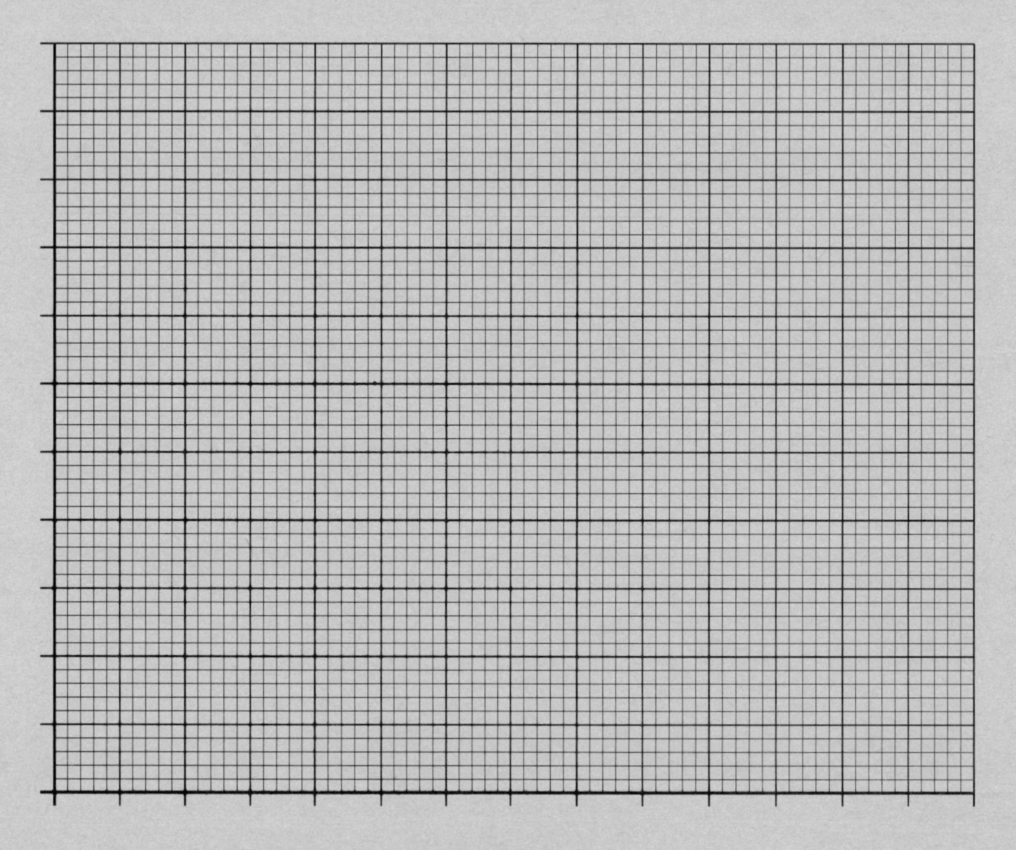

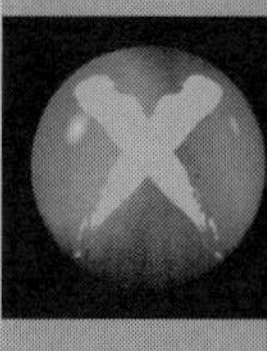

[BLANK PAGE]

FOR OFFICIAL USE

Total for Sections B & C

X009/301

NATIONAL QUALIFICATIONS 2002

FRIDAY, 31 MAY
1.00 PM – 3.30 PM

HUMAN BIOLOGY HIGHER

Fill in these boxes and read what is printed below.

Full name of centre

Town

Forename(s)

Surname

Date of birth
Day Month Year

Scottish candidate number

Number of seat

SECTION A—Questions 1–30

Instructions for completion of Section A are given on page two.

SECTIONS B AND C

1 (a) All questions should be attempted.

(b) It should be noted that in **Section C** questions 1 and 2 each contain a choice.

2 The questions may be answered in any order but all answers are to be written in the spaces provided in this answer book, and must be written clearly and legibly in ink.

3 Additional space for answers and rough work will be found at the end of the book. If further space is required, supplementary sheets may be obtained from the invigilator and should be inserted inside the **front** cover of this book.

4 The numbers of questions must be clearly inserted with any answers written in the additional space.

5 Rough work, if any should be necessary, should be written in this book and then scored through when the fair copy has been written.

6 Before leaving the examination room you must give this book to the invigilator. If you do not, you may lose all the marks for this paper.

SECTION A

Read carefully

1 Check that the answer sheet provided is for Human Biology Higher (Section A).

2 Fill in the details required on the answer sheet.

3 In this section a question is answered by indicating the choice A, B, C or D by a stroke made in **ink** in the appropriate place in the answer sheet—see the sample question below.

4 For each question there is only **one** correct answer.

5 Rough working, if required, should be done only on this question paper—or on the rough working sheet provided—**not** on the answer sheet.

6 At the end of the examination the answer sheet for Section A **must** be placed **inside** this answer book.

Sample Question

The digestive enzyme pepsin is most active in the

A mouth

B stomach

C duodenum

D pancreas.

The correct answer is **B**—stomach. A **heavy** vertical line should be drawn joining the two dots in the appropriate box in the column headed **B** as shown in the example on the answer sheet.

If, after you have recorded your answer, you decide that you have made an error and wish to make a change, you should cancel the original answer and put a vertical stroke in the box you now consider to be correct. Thus, if you want to change an answer D to an answer B, your answer sheet would look like this:

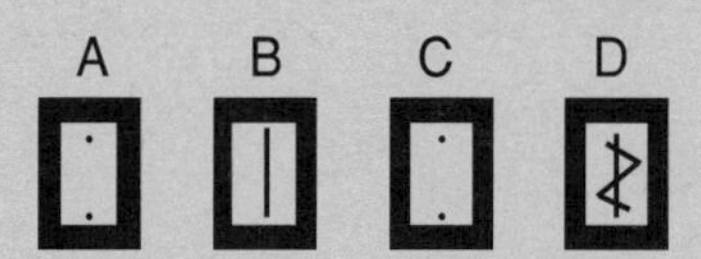

If you want to change back to an answer which has already been scored out, you should enter a tick (✓) to the **right** of the box of your choice, thus:

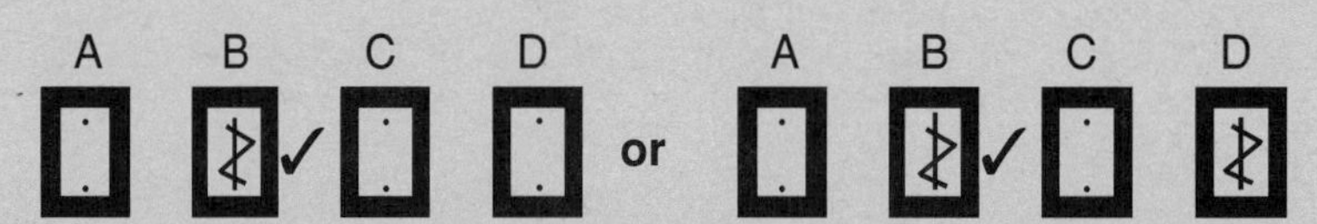

SECTION A

All questions in this section should be attempted.

Answers should be given on the separate answer sheet provided.

1. The cell membrane is chiefly composed of

A carbohydrates and lipids

B carbohydrates and proteins

C proteins and lipids

D proteins and nucleic acids.

2. Thirty percent of bases in a DNA molecule are adenine. The percentage of cytosine bases in the same molecule is

A 20%

B 30%

C 40%

D 70%.

3. Which of the following must be present for glycolysis to occur?

A Glucose and oxygen

B ATP and oxygen

C Glucose and ATP

D ATP and pyruvic acid

4. Glycolysis takes place in the

A nucleus

B cristae of mitochondria

C matrix of mitochondria

D cytoplasm.

5. The following diagram represents stages in the complete breakdown of glucose in aerobic respiration.

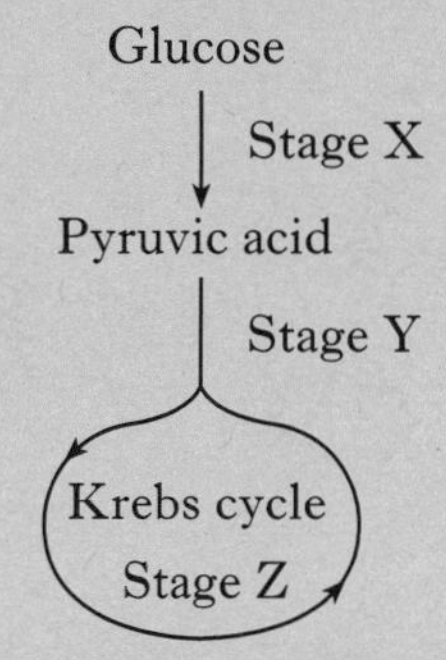

At which stage, or stages, is carbon dioxide released?

A Stages X and Z

B Stages X and Y

C Stages Y and Z

D Stage Z only

[Turn over

6. The diagram below shows part of a metabolic pathway. Each stage is controlled by an enzyme.

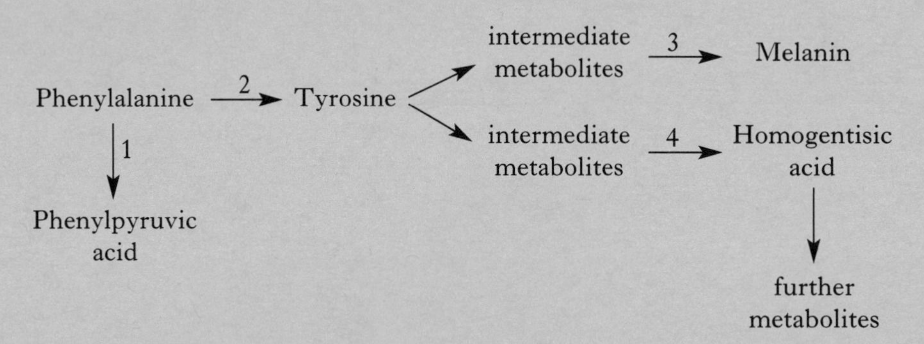

Phenylketonuria (PKU) is caused by a mutation of the gene required to make enzyme

A 1

B 2

C 3

D 4.

7. A stock solution has a concentration of 1 M. $100\,cm^3$ of a 0·6 M solution can be prepared by adding

A $40\,cm^3$ of stock solution to $60\,cm^3$ of water

B $60\,cm^3$ of stock solution to $40\,cm^3$ of water

C $60\,cm^3$ of stock solution to $100\,cm^3$ of water

D $100\,cm^3$ of stock solution to $60\,cm^3$ of water.

8. The table below shows some genotypes and phenotypes associated with forms of sickle-cell anaemia.

Genotype	*Phenotype*
Hb^AHb^A	normal
Hb^AHb^S	sickle-cell trait
Hb^SHb^S	acute sickle-cell anaemia

A normal man marries a woman with the sickle-cell trait. What are the chances that any child born to them will have acute sickle-cell anaemia?

A None

B 1 in 1

C 1 in 2

D 1 in 4

9. The diagram below shows the transmission of the gene for albinism.

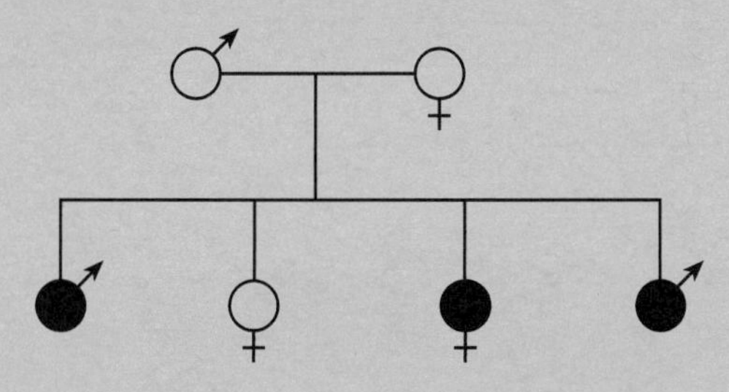

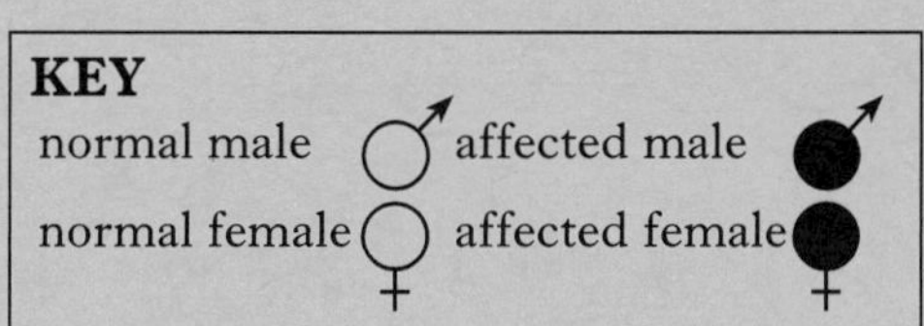

This condition is inherited as a characteristic which is

A dominant and not sex-linked

B recessive and not sex-linked

C dominant and sex-linked

D recessive and sex-linked.

10. Colour-blindness is a recessive, sex-linked characteristic controlled by the allele b.

Two parents with normal vision have a colour-blind boy.

The genotypes of the parents are

A X^BY and X^bX^b

B X^bY and X^BX^B

C X^bY and X^BX^b

D X^BY and X^BX^b.

11. The function of the interstitial cells in the human testes is to

A act as a store for sperm cells

B produce semen

C cause sperm cells to mature

D produce testosterone.

12. The diagram below shows a section through an ovary which contains developing eggs.

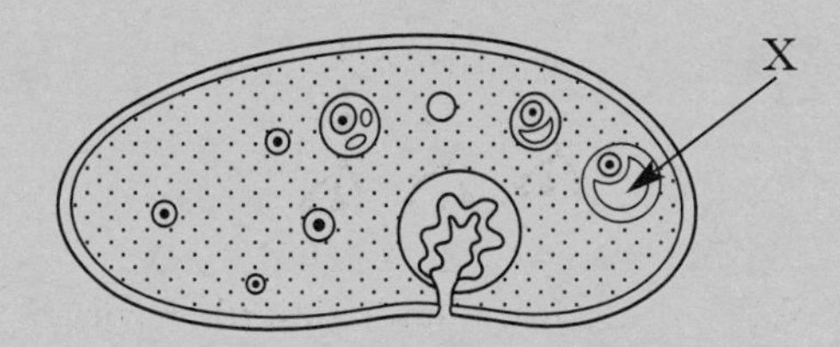

The structure labelled X is

A the endometrium

B a Graafian follicle

C the amnion

D a corpus luteum.

13. Which of the following materials are exchanged between maternal and fetal blood by diffusion?

A Oxygen and proteins

B Carbon dioxide and antibodies

C Oxygen and glucose

D Carbon dioxide and oxygen

14. The graph below represents an arterial blood pressure trace.

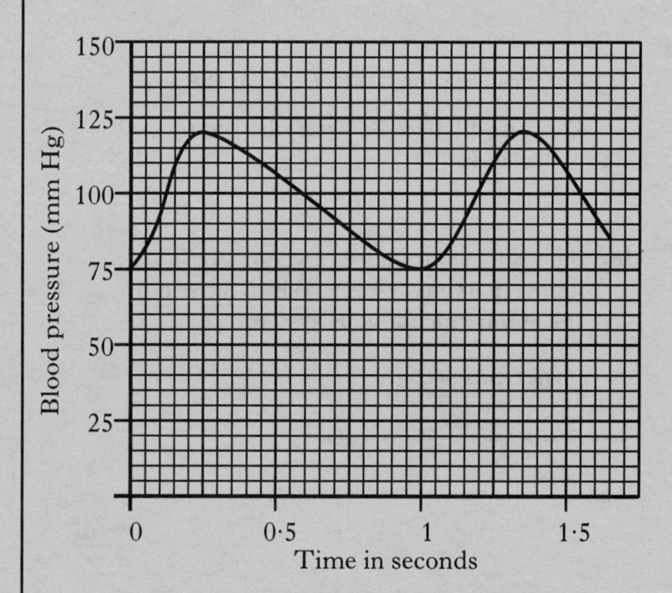

The blood pressure would be recorded as

A $\frac{120}{75}$

B $\frac{75}{120}$

C $\frac{104}{75}$

D $\frac{75}{104}$

15. The percentage distribution of blood groups in Scotland is shown below.

	Blood Group			
	O	A	B	AB
Scots (%)	51	35	11	3

What percentage of Scots could be given a blood transfusion of blood group A?

A 35%

B 38%

C 51%

D 86%

16. A person produces 1·5 litres of urine in 24 hours. The urine contains 36 g of urea.

What is the concentration of urea in the urine?

A 1·0 g/100 cm^3

B 2·4 g/litre

C 2·4 g/100 cm^3

D 3·6 g/100 cm^3

17. Stimulation of the sympathetic nerves causes

A vasoconstriction of arterioles in the skin

B vasoconstriction of the coronary arterioles

C vasodilation of arterioles of the gut

D vasodilation of arterioles in the salivary glands.

18. The flow chart shows the sub-divisions of the human nervous system. Which letter represents the autonomic nervous system?

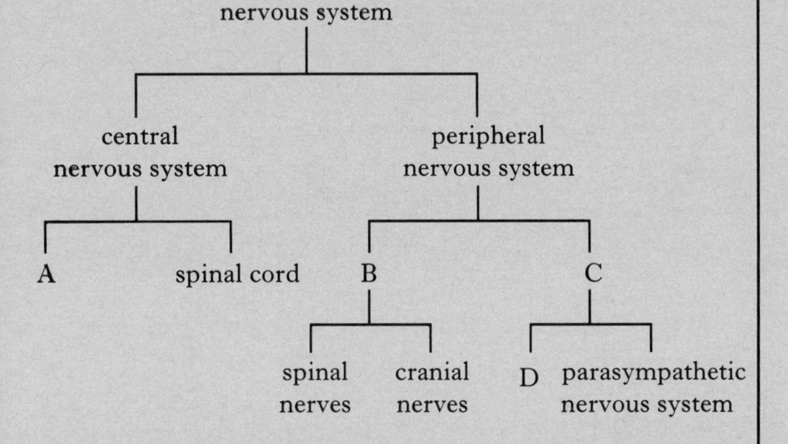

19. The somatic nervous system controls the

A skeletal muscles

B heart and blood vessels

C endocrine glands

D muscular wall of the gut.

20. The speed of impulse transmission along an axon is promoted by

A diffusion of neurotransmitters

B converging neural pathways

C diverging neural pathways

D myelination of fibres.

21. Which of the following statements about diverging neural pathways is correct?

A They accelerate the transmission of sensory impulses.

B They suppress the transmission of sensory impulses.

C They increase the degree of fine motor control.

D They decrease the degree of fine motor control.

22. The following diagram represents four neurones in a nervous pathway.

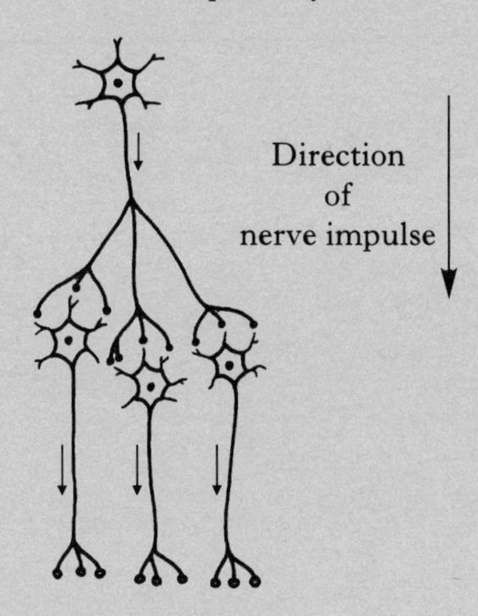

Which line of the table describes correctly the pathway?

	Type of Pathway	
A	sensory	convergent
B	motor	convergent
C	sensory	divergent
D	motor	divergent

23. The retrieval of information from long term memory is often aided by remembering the situation in which the information was encoded. This is described as using

A contextual cues

B chunking techniques

C rehearsal methods

D memory span.

24. A child, who was scratched by a black cat, now responds with a fear of all cats. This is an example of

A shaping

B reinforcement

C generalisation

D discrimination.

25. The diagram below shows the ages (in months) at which children reach various stages in their development.

The left end of each bar indicates the age by which 25% of infants have reached the stated performance.

The right end of each bar indicates the age by which 90% of infants have reached the stated performance.

The vertical bar indicates the age by which 50% of infants have reached the stated performance.

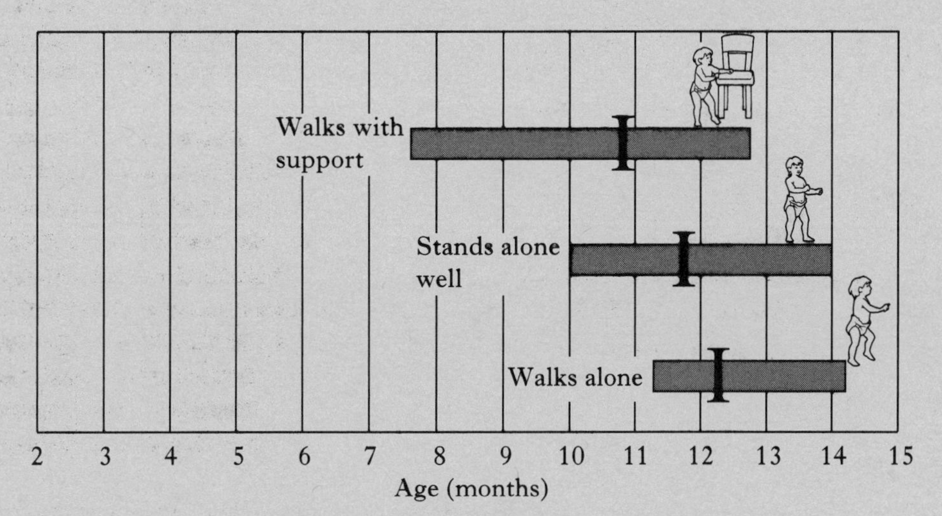

A nine-month-old infant can stand without support but cannot walk without support.

In what percentage of the population does this child lie?

A Less than 25%

B Around 25%

C Around 50%

D Greater than 50%

26. The diagram below shows the average lifespan of people in Britain between 1900 and 1990.

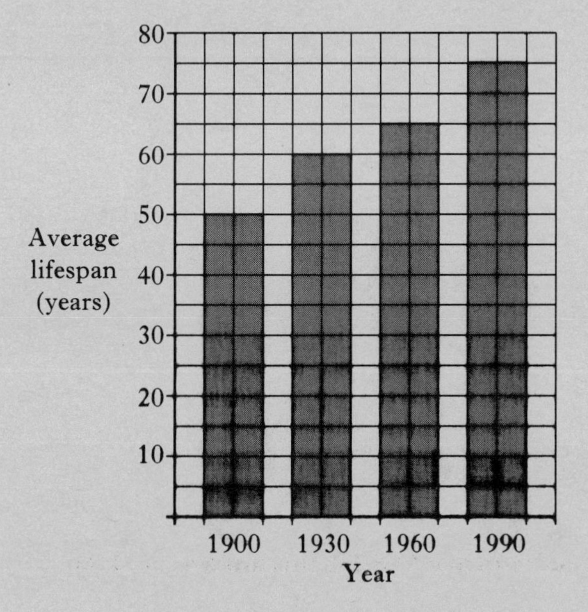

What is the percentage increase in lifespan during this period?

A 25%

B 45%

C 50%

D 75%

27. In the nitrogen cycle, which of the following processes is carried out by nitrifying bacteria?

The conversion of

A nitrate to ammonia

B ammonia to nitrate

C nitrogen gas to ammonia

D nitrogen gas to nitrate.

28. In a river, samples of water from above and below a sewage outlet were compared. Which comparison is correct?

	Sample above sewage outlet	*Sample below sewage outlet*
A	high oxygen concentration	many bacteria
B	high oxygen concentration	few bacteria
C	low oxygen concentration	many bacteria
D	low oxygen concentration	few bacteria

29. The graphs below contain information about the population of Britain.

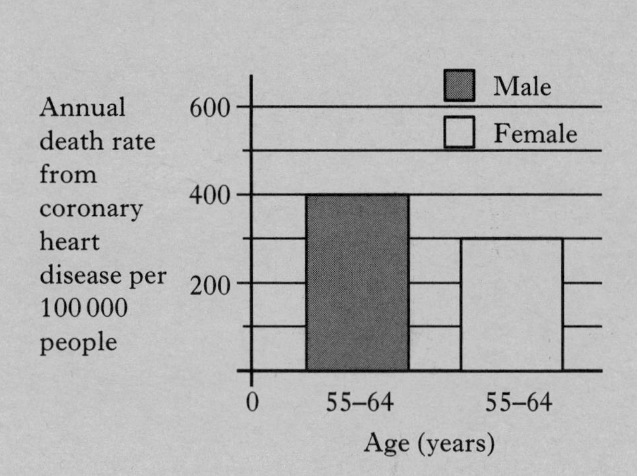

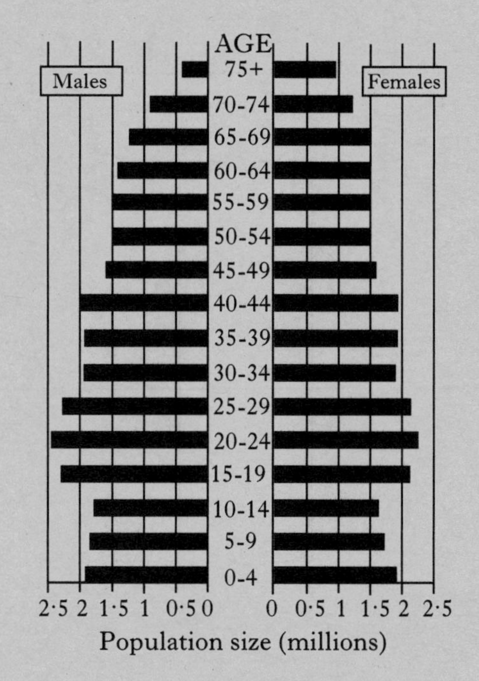

How many British women between 55 and 64 years of age die from coronary heart disease annually?

A 300

B 4500

C 9000

D 21 000

30. Which of the following pairs of gases are the principal contributors to the greenhouse effect?

A Nitrogen and carbon dioxide

B Carbon dioxide and methane

C Ammonia and carbon dioxide

D Nitrogen and methane

Candidates are reminded that the answer sheet MUST be returned INSIDE this answer booklet.

[Turn over for Section B]

DO NOT WRITE IN THIS MARGIN

Marks

SECTION B

All questions in this section should be attempted.

1. The diagram below shows a section through a nucleus and associated cell structures.

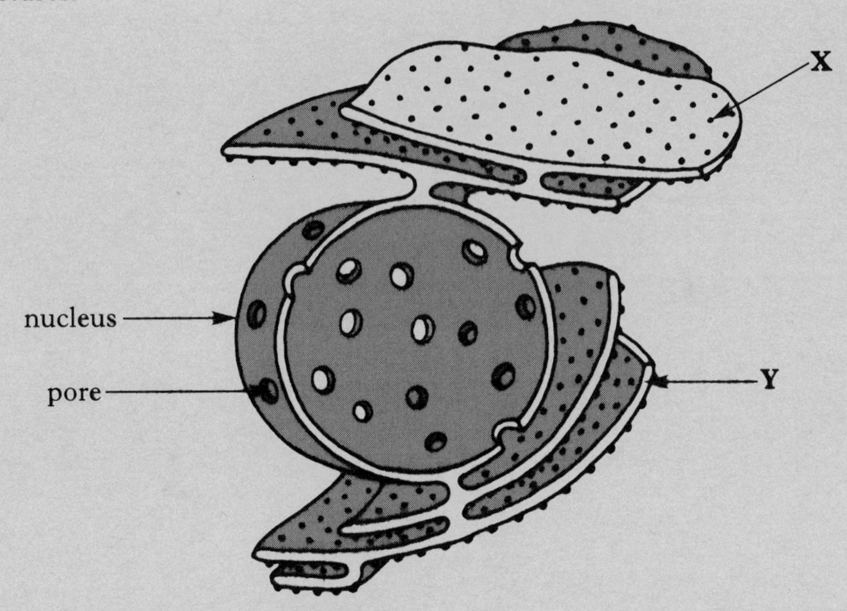

(*a*) (i) Name organelle **X**.

_______________ 1

(ii) What type of substance is manufactured by organelle **X**?

_______________ 1

(iii) Give an example of such a substance.

_______________ 1

(*b*) (i) The structure labelled **Y** is composed of sheets of membranes.

What name is given to this structure?

_______________ 1

(ii) Structure **Y** can transport substances to another organelle within the cell.

Give an example of such an organelle and state its function.

Example _______________ 1

Function _______________ 1

(*c*) Why is it necessary to have pores in the nuclear membrane?

____

_______________ 1

DO NOT WRITE IN THIS MARGIN

Marks

2. The diagram shows the synthesis of a peptide chain.

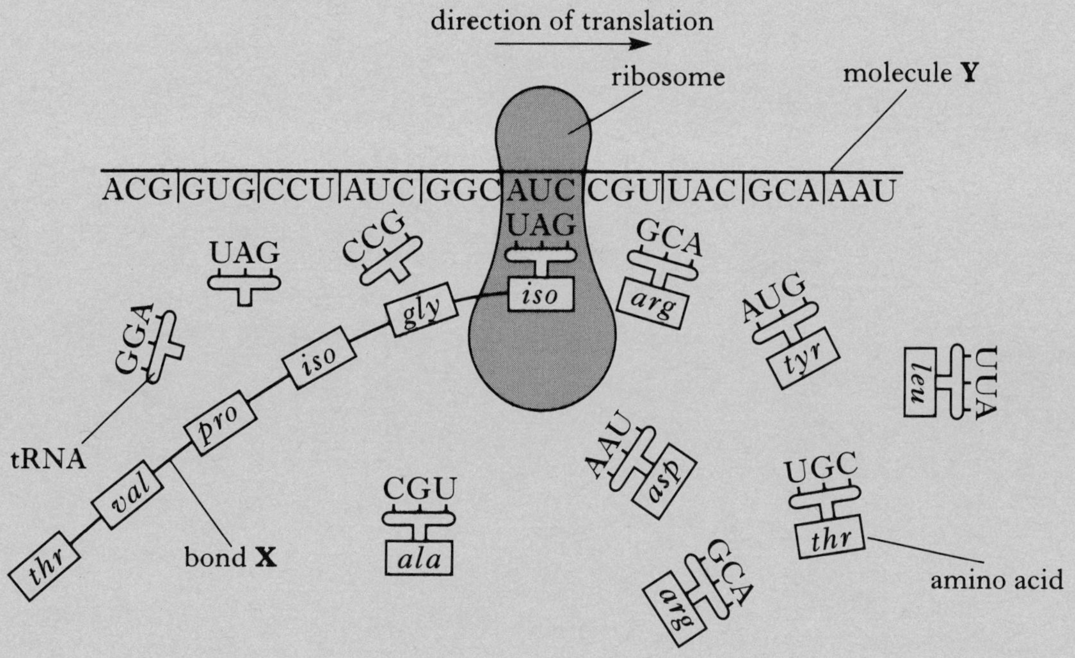

(*a*) Name bond **X** and molecule **Y**.

Bond **X** ________________ Molecule **Y** ________________ **2**

(*b*) What term is used to describe the triplet code on the tRNA molecules?

________________ **1**

(*c*) Give the abbreviated names of the next **four** amino acids which will be attached to complete the peptide chain.

iso → ________ → ________ → ________ → ________ **1**

(*d*) What sequence of bases on a DNA molecule will code for the amino acid labelled *thr*?

________________ **1**

(*e*) Amino acids are added to the peptide chain at the rate of 15 per second.

How long will it take for the complete synthesis of the peptide shown in the diagram above?

________ s **1**

[Turn over

DO NOT WRITE IN THIS MARGIN

Marks

3. The diagram below shows the inheritance of tongue-rolling ability in a family.

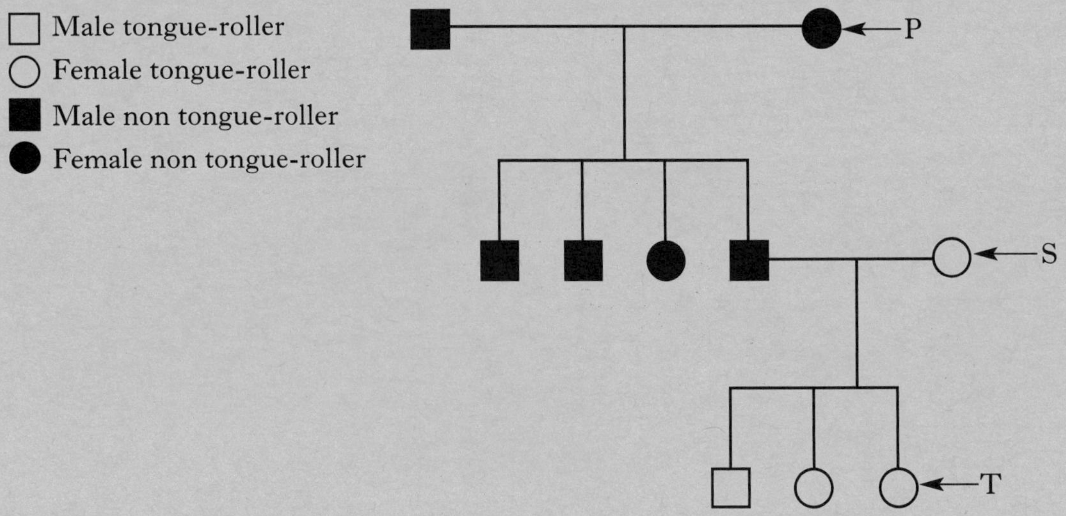

□ Male tongue-roller
○ Female tongue-roller
■ Male non tongue-roller
● Female non tongue-roller

P

S

T

(*a*) (i) Using the symbol **R** to represent the allele for tongue-rolling, and the symbol **r** to represent the allele for non tongue-rolling, state the genotypes of individuals P and T.

P ________ T ________ **1**

(ii) How many individuals, shown in this family tree, have a genotype which is homozygous recessive?

________ **1**

(iii) Place a cross through a symbol in the family tree which represents a heterozygous male. **1**

(iv) Female S is pregnant. Using information from the family tree, is it possible to predict whether the child will be a tongue-roller or not? Give a reason for your answer.

YES/NO ________

Reason ________________________________

________________________________ **1**

(*b*) Some characteristics are controlled by several genes.

(i) State the term used to describe this type of inheritance pattern.

________________________________ **1**

(ii) Which **two** of the following human characteristics show this type of inheritance pattern? *Underline the correct answers.*

blood groups **cystic fibrosis** **height**

haemophilia **skin colour** **1**

DO NOT WRITE IN THIS MARGIN

Marks

4. Some stages in the process of meiosis are shown in the diagrams below.

Only 6 chromosomes are shown in each cell.

A 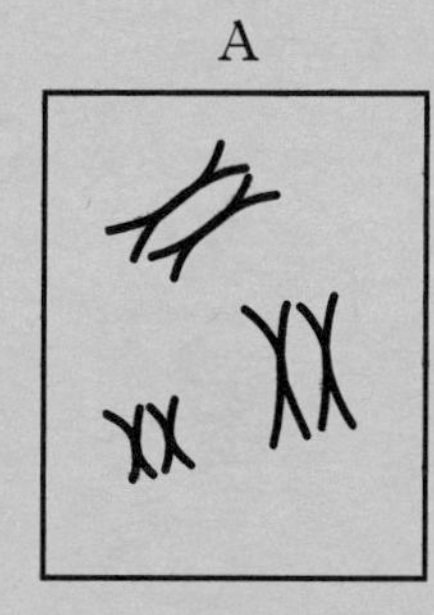

B 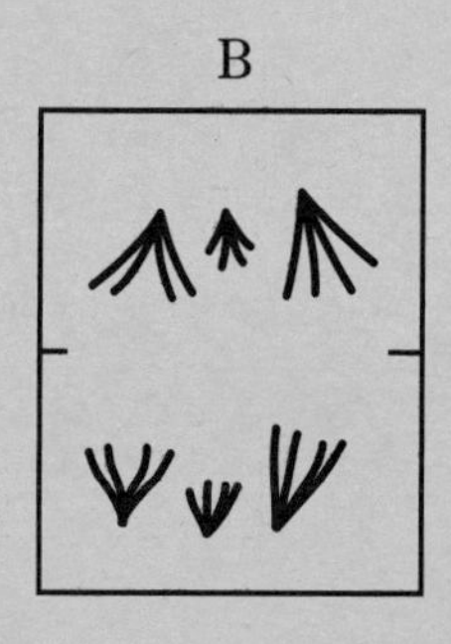

C 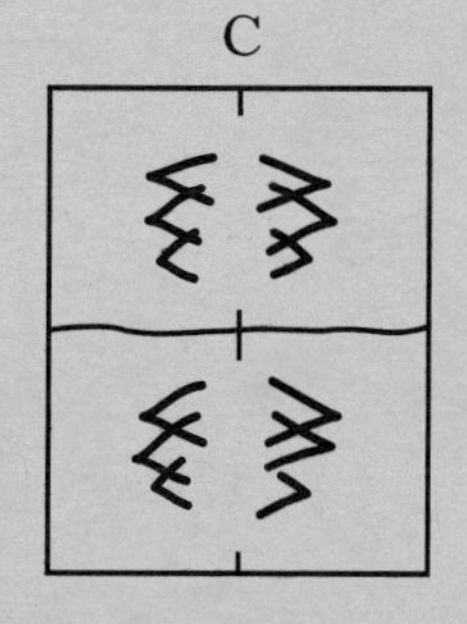

D 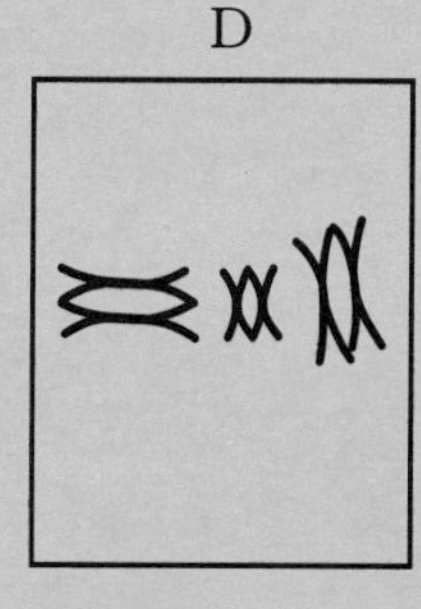

(*a*) (i) Place the stages in the correct order.

________ → ________ → ________ → ________ **1**

(ii) At which stage could chiasmata form?

________ **1**

(iii) Why are chiasmata important?

____

____ **1**

(*b*) The diagram below shows a gamete mother cell and four sperm which would result from meiosis.

(i) Complete the diagram by writing in the normal number of chromosomes found in each of the cells. **1**

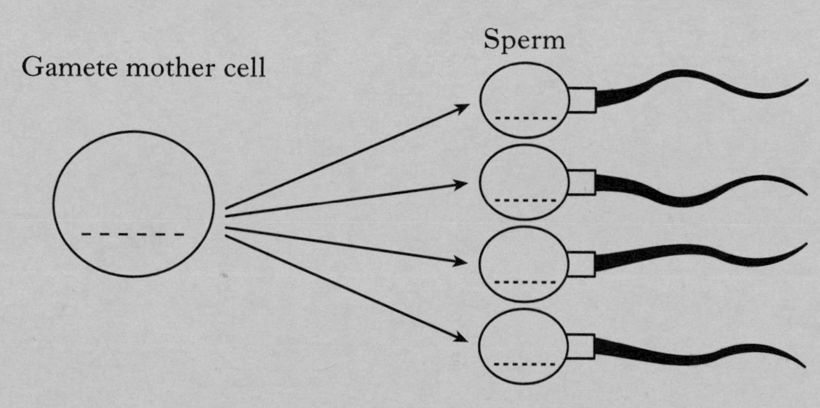

(ii) In how many of these sperm will an **X** chromosome be found?

________ **1**

(iii) Where in the testes does meiosis occur?

____________________ **1**

[Turn over

DO NOT WRITE IN THIS MARGIN

Marks

5. The graph below shows the relative concentrations of three hormones in the plasma of a woman during a normal 28-day menstrual cycle.

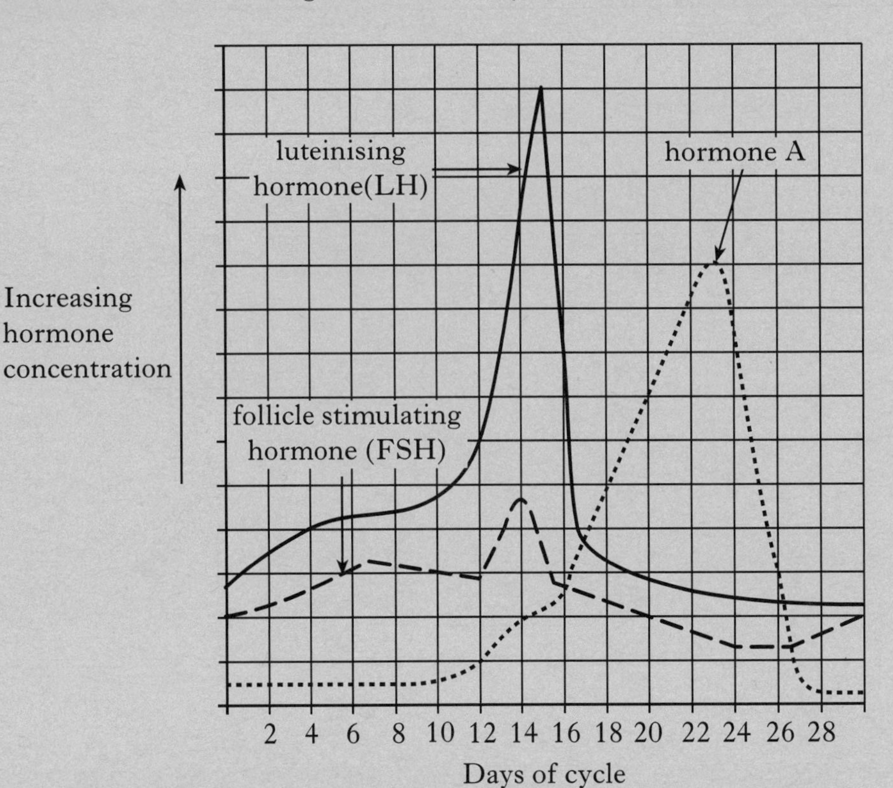

(*a*) Name hormone A.

____________________ **1**

(*b*) What is the effect of the sudden increase in concentration of luteinising hormone?

____________________ **1**

(*c*) During which time period is the endometrium likely to reach maximum thickness?

Underline the correct answer.

0–4 days **12–16 days** **22–26 days** **1**

(*d*) In what way would the line showing the concentration of FSH be different if fertilisation took place during this cycle? Give an explanation for your answer.

Difference ____________________

____________________ **1**

Explanation ____________________

____________________ **1**

DO NOT WRITE IN THIS MARGIN

Marks

6. The diagram below shows the heart and its associated nerves.

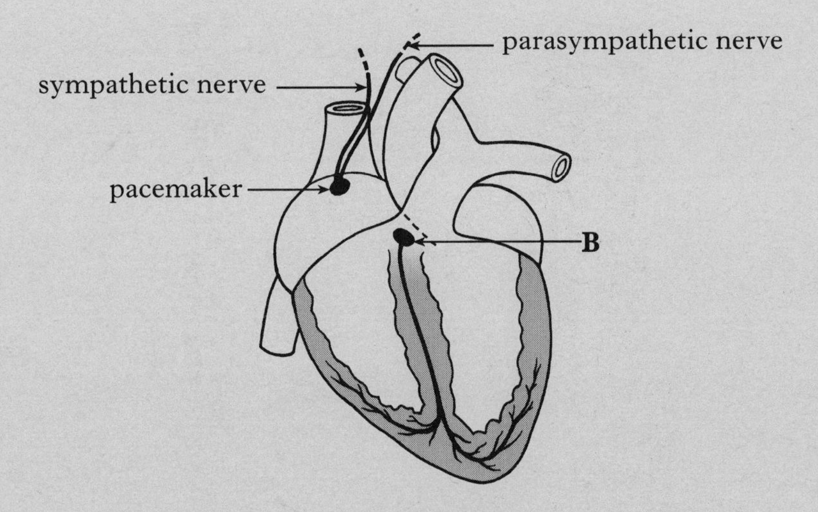

(*a*) On the diagram, mark with an **X** the chamber where the blood pressure is highest during the cardiac cycle. **1**

(*b*) Describe the effect of impulses from the parasympathetic nerve on the heart.

______________________________ **1**

(*c*) (i) Name the part of the heart labelled **B**.

______________________________ **1**

(ii) Describe the role of **B** in the cardiac cycle.

____

______________________________ **1**

(iii) An individual has a heart rate of 75 bpm. How long does one cardiac cycle last?

Space for calculation

______________ s **1**

[Turn over

DO NOT WRITE IN THIS MARGIN

Marks

7. The diagram below shows the structure of a villus.

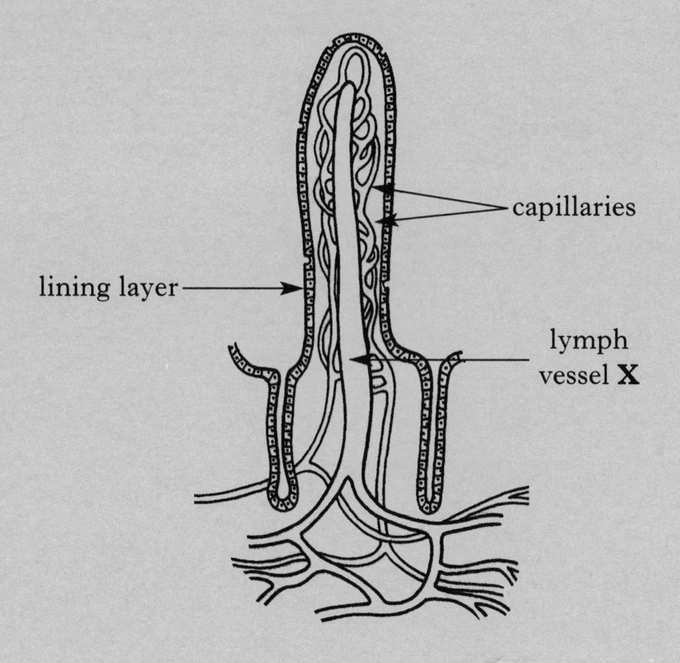

(*a*) (i) Name lymph vessel **X**. **1**

____

(ii) Describe the role of lymph vessel **X** in the transport of nutrients. **1**

____

____

(*b*) (i) Vitamin B_{12} is absorbed into the blood capillaries of the villus.

Name the substance which must be present before vitamin B_{12} can be absorbed. **1**

____

(ii) Which type of body cell requires vitamin B_{12} for its manufacture? **1**

____

(*c*) Name the blood vessel which transports nutrient-rich blood away from the small intestine. **1**

____

DO NOT WRITE IN THIS MARGIN

Marks

8. The diagram below shows blood flow through a capillary bed in the pancreas. The area shown is an Islet of Langerhans.

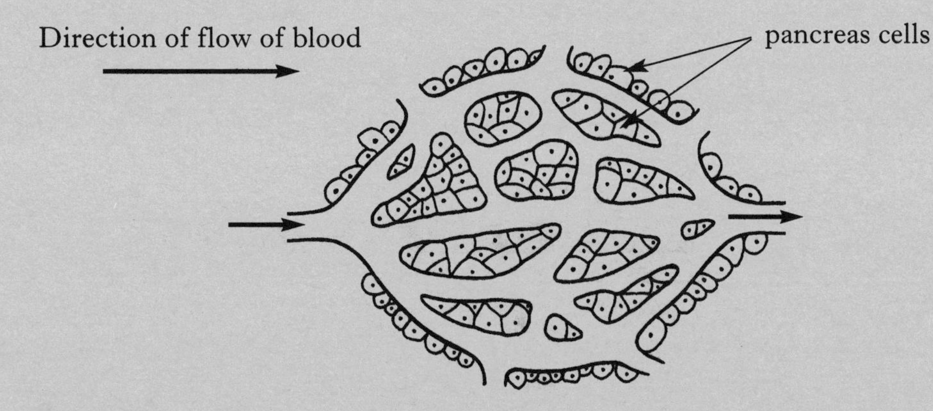

(*a*) After a meal rich in carbohydrate, the composition of the blood changes as it flows through the pancreas.

Complete the table below using the words *increases* or *decreases* to indicate these changes.

Substance	*Change*
Oxygen	
Carbon dioxide	
Insulin	

2

(*b*) (i) What effect does insulin have on the body?

____

____ **1**

(ii) Name a hormone produced by the pancreas which has the opposite effect to insulin.

________________________ **1**

[Turn over

9. The graphs below contain information about diet in the UK.

Graph 1 shows how UK diet has changed between 1988 and 1998.

Graph 2 shows how Scottish consumption compares with the rest of the UK in 1998.

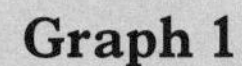

Graph 1

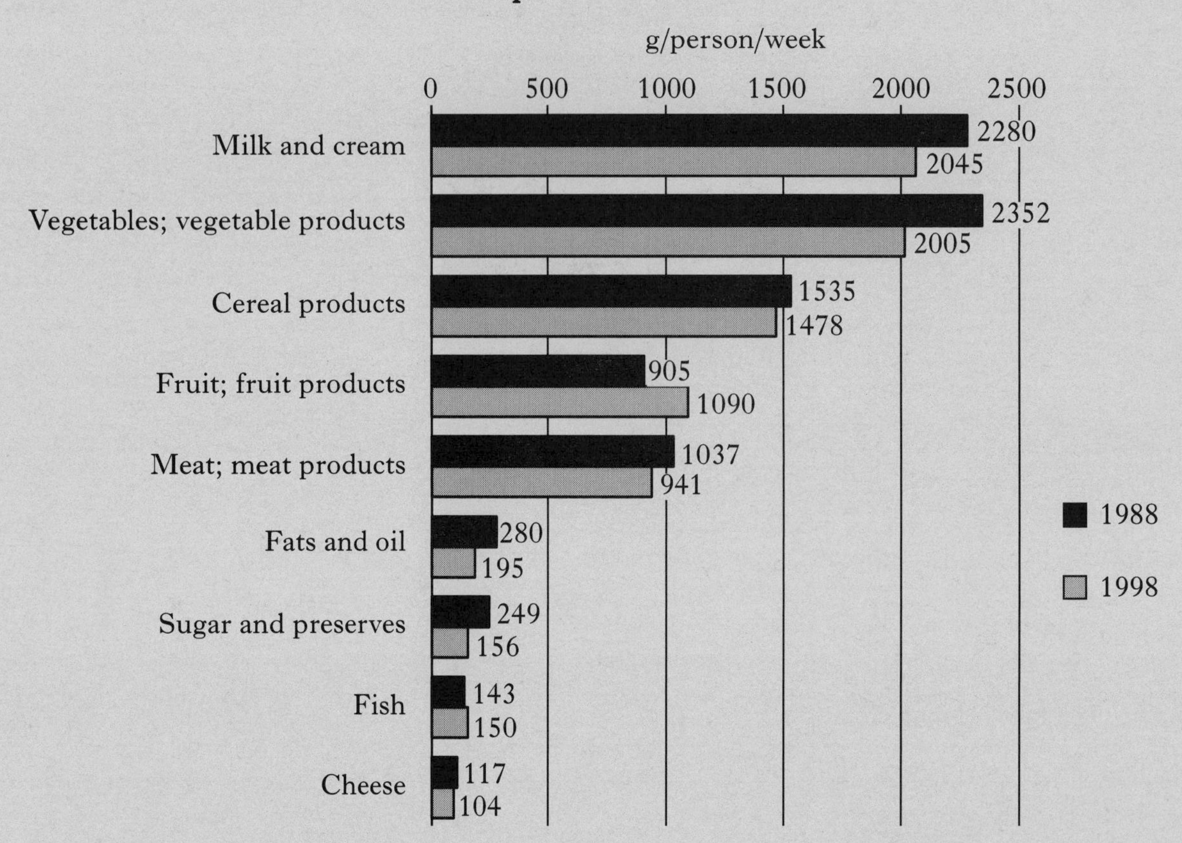

Graph 2

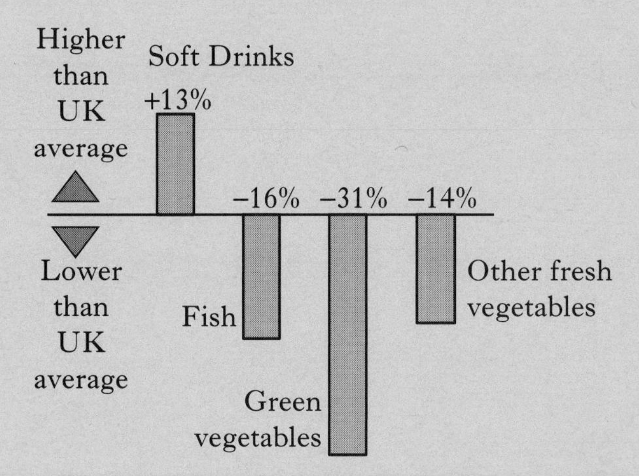

DO NOT WRITE IN THIS MARGIN

Marks

9. (continued)

(*a*) (i) From **Graph 1**, what general conclusion can be drawn about UK diet in 1988 compared with 1998?

____ **1**

(ii) What additional information would be required in order to draw a general conclusion about the Scottish diet over the same period of time?

____ **1**

(iii) When collecting data to make comparisons of this type, state **two** variables which should be controlled.

1 ____

2 ____ **2**

(*b*) From **Graph 1**, calculate the percentage reduction in milk and cream consumption over the decade.

Space for calculation

________________ % **1**

(*c*) With reference to **both** graphs, calculate the weekly fish consumption in Scotland in 1998.

Space for calculation

________________ g/person/week **1**

(*d*) The incidence of coronary heart disease in the UK over the ten-year period has decreased. From **Graph 1** give **two** pieces of evidence which may contribute to this decrease.

1 ____

2 ____ **1**

(*e*) Information such as this can be used by governments to plan for the future. Suggest what use might be made of the information on diet shown in the graphs.

____

____

____

____ **1**

DO NOT WRITE IN THIS MARGIN

Marks

10. The diagram shows a section through the brain.

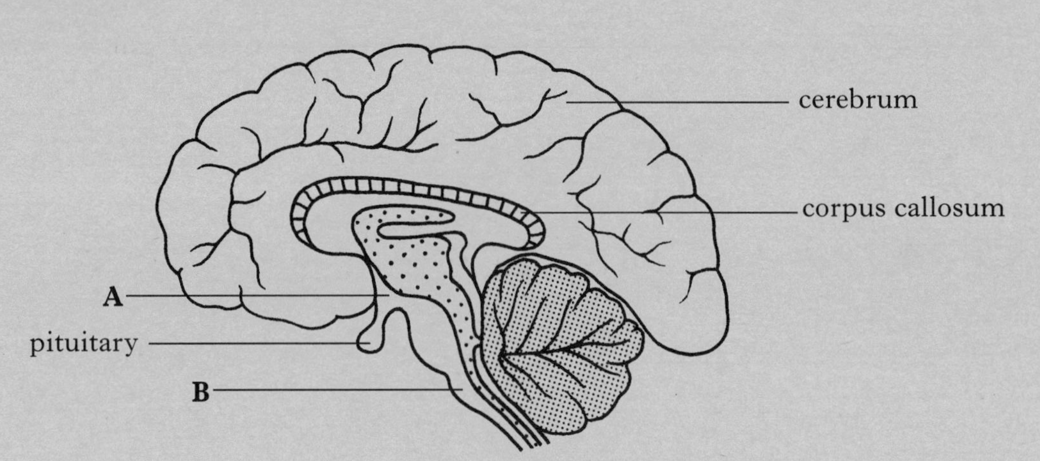

(*a*) Name parts **A** and **B** shown on the diagram.

A ______________________ **B** ______________________ **2**

(*b*) (i) Name **two** areas of the cerebrum in which functions are localised.

______________________ and ______________________ **1**

(ii) Explain how the convoluted surface of the cerebrum contributes to its function.

____

____ **1**

(iii) What is the function of the corpus callosum?

____

____ **1**

(*c*) Parts of the brain are involved in memory storage. Complete the following sentences which relate to memory loss, using words from the list below.

Alzheimer's	**noradrenaline**	**limbic**
Huntingdon's	**acetylcholine**	**lymphatic**

A disorder particularly associated with memory loss is ______________

disease. This disorder is due to the disappearance of cells which produce the

neurotransmitter______________ in the______________ system

of the brain. **2**

DO NOT WRITE IN THIS MARGIN

Marks

10. (continued)

(*d*) The hormone ADH is produced by the pituitary gland.

Describe the role of ADH in restoring water balance after excess water has been lost from the body.

____

____

____

____

____ **3**

[Turn over

DO NOT WRITE IN THIS MARGIN

Marks

11. A student carried out an experiment to investigate how quickly three groups of pupils completed a finger maze.

The table below shows the fastest times obtained by each group of pupils.

	Fastest time achieved in trials (s)		
	8 year-olds	*12 year-olds*	*16 year-olds*
	11	7	7
	12	8	8
	9	6	7
	6	10	9
	11	8	6
	10	7	–
	12	12	–
	13	6	–
	–	7	–
	–	6	–
average	**10·5**		**7·4**

(*a*) Calculate the average time for the 12 year-old pupils and write your answer in the table above. **1**

(*b*) Construct a **bar graph** to show the **average** times for each age group of pupils.

(Additional graph paper, if required, will be found on page 30.)

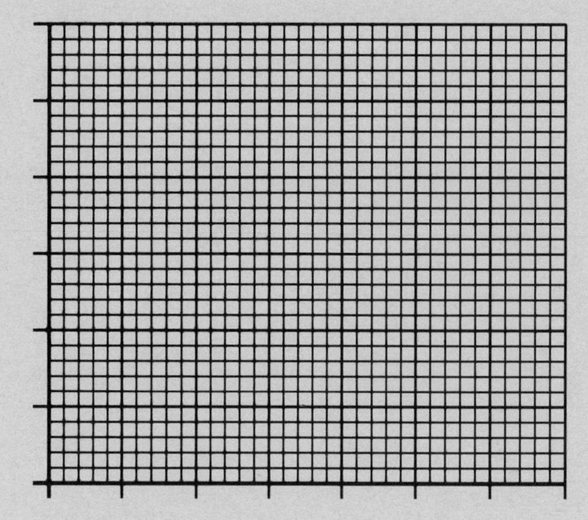

2

DO NOT WRITE IN THIS MARGIN

Marks

11. (continued)

(*c*) (i) Explain why these average results are not reliable.

____

____ **1**

(ii) Identify another factor which would have to be considered in the design of this experiment to ensure that valid conclusions can be drawn.

____

____ **1**

(*d*) It was found that some of the pupils performed better when they were being watched by other pupils. What term is used to describe this effect?

______________________ **1**

[Turn over

DO NOT WRITE IN THIS MARGIN

Marks

12. The graph below shows two survival curves for the UK population.

The curves plot the number of people at each age who are still alive.

One curve is from the year 1800 and the other from the year 1960.

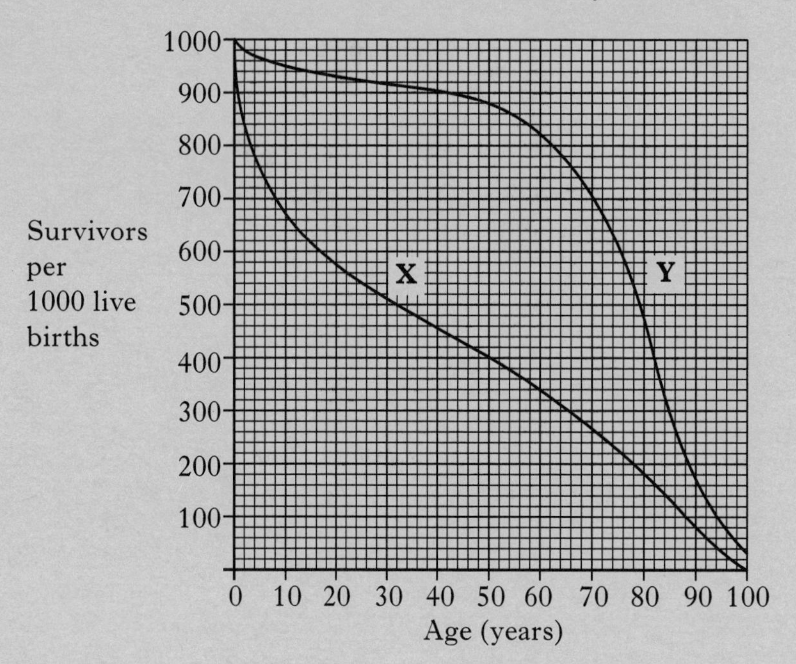

(*a*) Complete the table to identify each curve and give a reason for your selection.

Curve	*Year*
X	
Y	

Reason for selection

____

____ **1**

(*b*) Suggest **two** reasons for the difference in the curves between the ages of 0 and 10.

1 ____

____

2 ____

____ **1**

DO NOT WRITE IN THIS MARGIN

12. (continued) *Marks*

(*c*) For each curve, express the numbers still alive at the age of 50 as a percentage of their original population.

Space for calculation

X ________________ % **Y** ________________ % **1**

(*d*) For each of the curves identify the ten-year period during which there is the highest mortality rate.

X ________________ years **Y** ________________ years **1**

[Turn over

13. The graphs below relate to fertiliser use by farmers, and nitrate pollution of two rivers from 1930 to 1990.

Graph 1. The use of nitrogen (N) and phosphorus (P) in fertiliser.

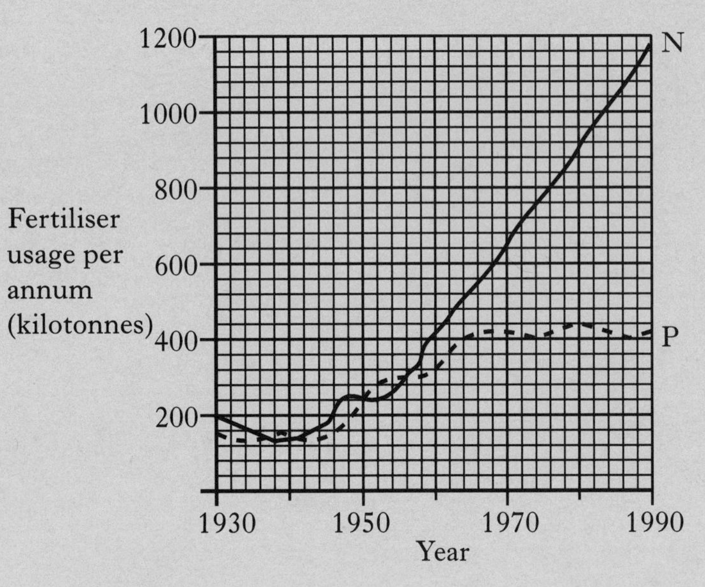

Graph 2. The nitrate content of the rivers Dee and Tay.

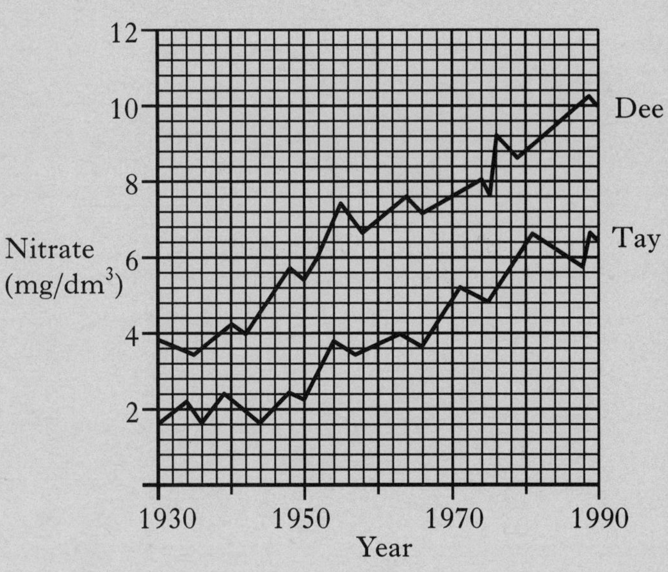

N – Nitrogen
P – Phosphorus

DO NOT WRITE IN THIS MARGIN

Marks

(*a*) From **Graph 1**, compare the change in use of nitrogen and phosphorus in fertiliser from 1930 to 1990.

____

____

____

____ **2**

(*b*) From **Graph 2**, what was the nitrate concentration in the river Dee in the year 1970?

______________ **1**

(*c*) What evidence is there, from comparison of both graphs, that agricultural use of fertiliser is linked to nitrate pollution of the Dee and Tay?

____

____ **1**

DO NOT WRITE IN THIS MARGIN

Marks

13. (continued)

(*d*) Describe **two** effects of increased nitrate pollution of fresh water.

1 ____

____

2 ____

____ **2**

(*e*) Name a source of nitrate and phosphate pollution other than farmland.

______________________ **1**

(*f*) Describe **one** way in which phosphorus is important to the structure or function of a cell.

____

____ **1**

[Turn over

Marks

SECTION C

Both questions in this section should be attempted.

Note that each question contains a choice.

Questions 1 and 2 should be attempted on the blank pages which follow.

Supplementary sheets, if required, may be obtained from the invigilator.

Labelled diagrams may be used where appropriate.

1. Answer **either** A **or** B.

A. Give an account of DNA under the following headings:

(i) DNA structure; **7**

(ii) DNA replication. **3**

(10)

OR

B. Give an account of immunity under the following headings:

(i) B-lymphocytes and T-lymphocytes; **7**

(ii) Macrophages. **3**

(10)

In question 2 ONE mark is available for coherence and ONE mark is available for relevance.

2. Answer **either** A **or** B.

A. Give an account of the life history of a red blood cell. **(10)**

OR

B. Give an account of the influence of hormones on the growth and development of boys. **(10)**

[*END OF QUESTION PAPER*]

DO NOT WRITE IN THIS MARGIN

SPACE FOR ANSWERS

DO NOT WRITE IN THIS MARGIN

SPACE FOR ANSWERS

ADDITIONAL GRAPH PAPER FOR QUESTION 11(*b*)

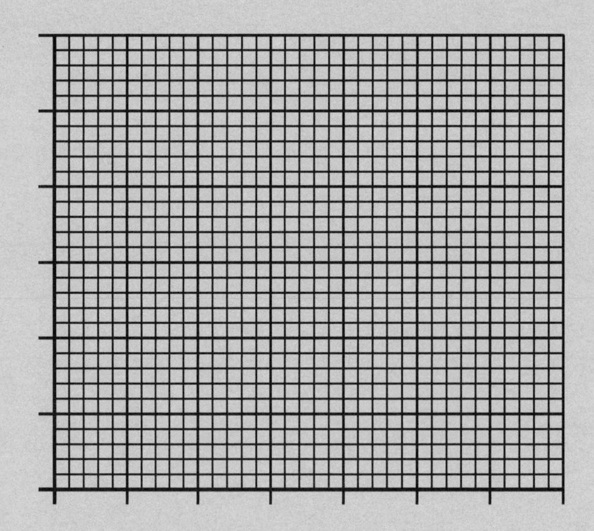

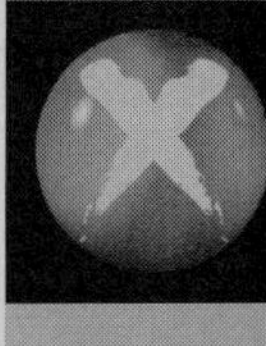

[BLANK PAGE]

FOR OFFICIAL USE

Total for Sections B & C

X009/301

NATIONAL QUALIFICATIONS 2003

MONDAY, 26 MAY
1.00 PM – 3.30 PM

HUMAN BIOLOGY HIGHER

Fill in these boxes and read what is printed below.

Full name of centre

Town

Forename(s)

Surname

Date of birth
Day Month Year

Scottish candidate number

Number of seat

SECTION A—Questions 1–30

Instructions for completion of Section A are given on page two.

SECTIONS B AND C

1 (a) All questions should be attempted.

(b) It should be noted that in **Section C** questions 1 and 2 each contain a choice.

2 The questions may be answered in any order but all answers are to be written in the spaces provided in this answer book, and must be written clearly and legibly in ink.

3 Additional space for answers and rough work will be found at the end of the book. If further space is required, supplementary sheets may be obtained from the invigilator and should be inserted inside the **front** cover of this book.

4 The numbers of questions must be clearly inserted with any answers written in the additional space.

5 Rough work, if any should be necessary, should be written in this book and then scored through when the fair copy has been written.

6 Before leaving the examination room you must give this book to the invigilator. If you do not, you may lose all the marks for this paper.

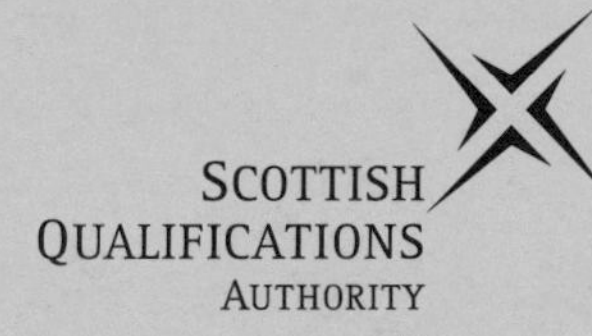

SECTION A

Read carefully

1 Check that the answer sheet provided is for Human Biology Higher (Section A).

2 Fill in the details required on the answer sheet.

3 In this section a question is answered by indicating the choice A, B, C or D by a stroke made in **ink** in the appropriate place in the answer sheet—see the sample question below.

4 For each question there is only **one** correct answer.

5 Rough working, if required, should be done only on this question paper—or on the rough working sheet provided—**not** on the answer sheet.

6 At the end of the examination the answer sheet for Section A **must** be placed **inside** this answer book.

Sample Question

The digestive enzyme pepsin is most active in the

A mouth

B stomach

C duodenum

D pancreas.

The correct answer is **B**—stomach. A **heavy** vertical line should be drawn joining the two dots in the appropriate box in the column headed **B** as shown in the example on the answer sheet.

If, after you have recorded your answer, you decide that you have made an error and wish to make a change, you should cancel the original answer and put a vertical stroke in the box you now consider to be correct. Thus, if you want to change an answer D to an answer B, your answer sheet would look like this:

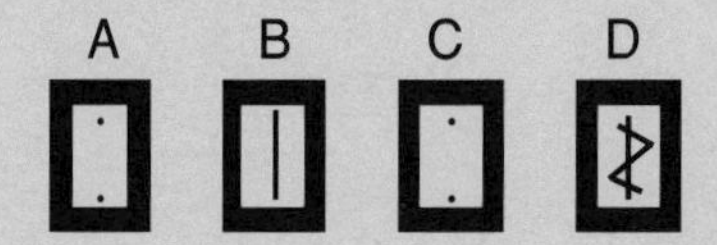

If you want to change back to an answer which has already been scored out, you should enter a tick (✓) to the **right** of the box of your choice, thus:

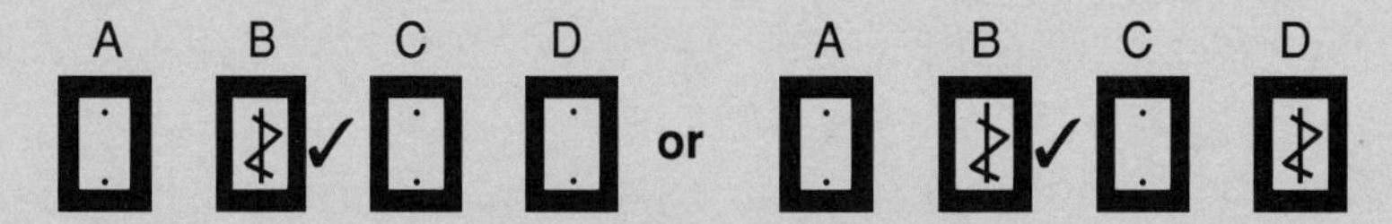

SECTION A

All questions in this section should be attempted.

Answers should be given on the separate answer sheet provided.

1. The diagram represents part of a molecule of DNA on which a molecule of RNA is being synthesised.

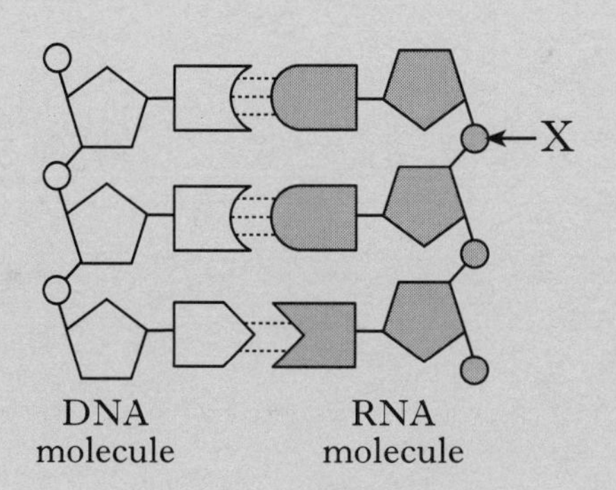

What does component X represent?

A Ribose sugar

B Deoxyribose sugar

C Phosphate

D Ribose phosphate

2. DNA controls the activities of a cell by coding for the production of

A proteins

B carbohydrates

C amino acids

D bases.

3. The diagram shows a stage in the synthesis of part of a polypeptide.

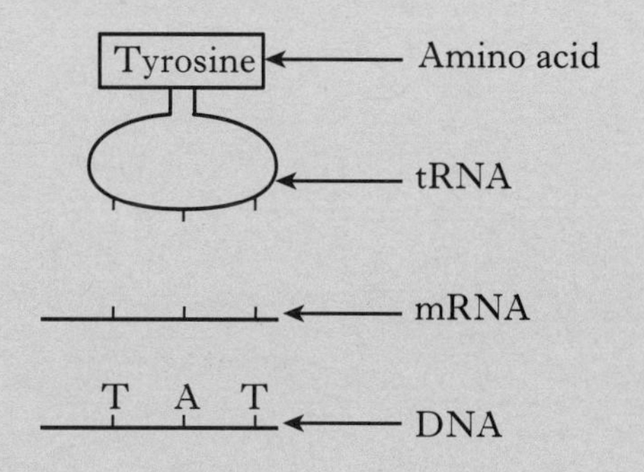

Identify the triplet codes for the amino acid tyrosine.

	mRNA	*tRNA*
A	ATA	UAU
B	UAU	AUA
C	AUA	UAU
D	ATA	TAT

4. The base sequence of a short piece of DNA is shown below.

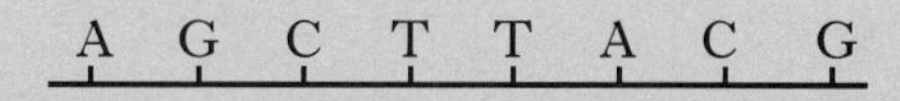

During replication, an inversion mutation occurred on the complementary strand synthesised on this piece of DNA.

Which of the following is the mutated complementary strand?

A T C G A A T G A

B A G C T T A G C

C T C G A A T C G

D T C G A A T G C

5. The graph shows the effect of substrate concentration on the rate of an enzyme-controlled reaction.

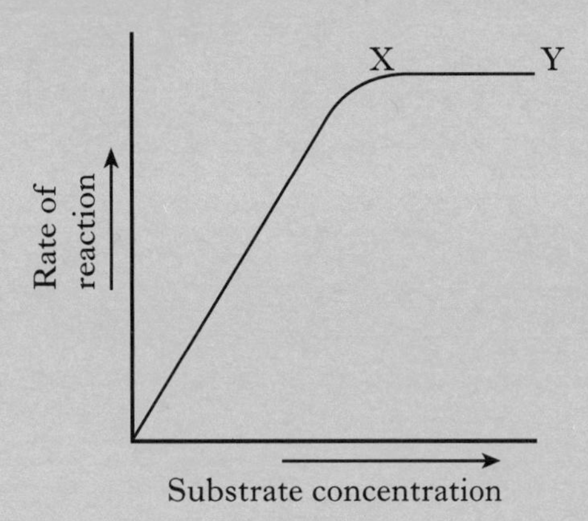

The graph levels out between points X and Y because the

A enzyme is denatured

B active sites are saturated with substrate

C enzyme is inhibited

D enzyme is activated.

6. When a protease enzyme is added to an amylase solution, which of the following could be produced?

A Amino acids

B Maltose

C Glucose

D Glycerol

7. The diagram below represents stages in tissue respiration.

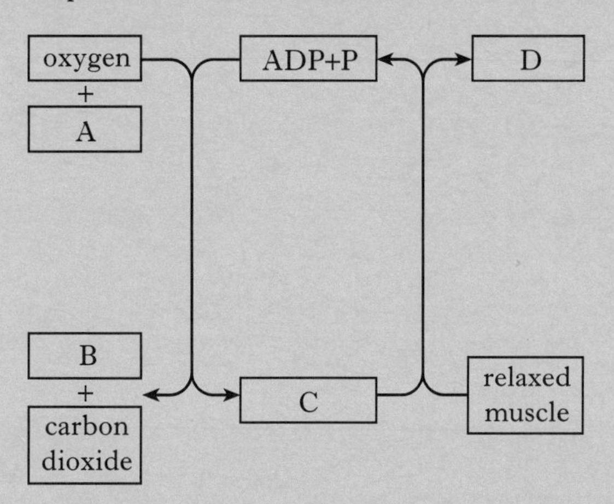

Which box represents ATP?

8. A piece of muscle was cut into three strips X, Y and Z and treated as described in the table.

Their final lengths were then measured.

Muscle strip	*Solution added to muscle*	*Muscle length* (mm)	
		Start	*After 10 minutes*
X	1% glucose	50	50
Y	1% ATP	50	45
Z	1% ATP boiled and cooled	50	46

From the data it may be deduced that

A ATP is not an enzyme

B muscles contain many mitochondria

C muscles synthesise ATP in the absence of glucose

D muscles do not use glucose as a source of energy.

9. Which line in the table has pairs of statements which are true with regard to aerobic respiration and anaerobic respiration in human muscle tissue?

	Aerobic respiration	*Anaerobic respiration*
A	There is a net gain of ATP	Carbon dioxide is not produced
B	There is a net gain of ATP	Oxygen is used up
C	Carbon dioxide is evolved	There is a net loss of ATP
D	Lactic acid is formed	Ethanol is formed

10. The graph below shows changes which occur in the masses of protein, fat and carbohydrate in a boy's body during five weeks of starvation.

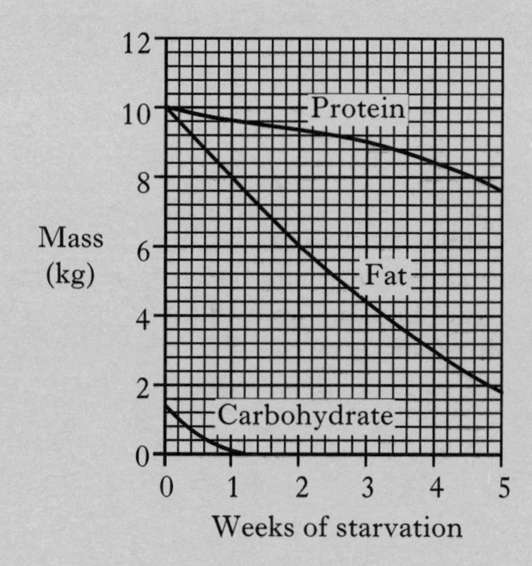

The boy weighs 60 kg. Predict his weight after two weeks without food.

A 43 kg

B 50 kg

C 54 kg

D 57 kg

11. Which of the following will result in the gain of active immunity by the body?

A Transfer of antibodies across the placenta

B Injection of antitoxin

C Suckling of breast milk

D Invasion by viruses

12. Which of the following reactions describes autoimmunity?

A The production of antibodies in response to infection

B The rapid production of antibodies in response to reinfection

C The production of antibodies in response to immunisation

D The production of antibodies in response to the body's own cells

13. Two parents, one of blood group A and the other of blood group B, have four children.

The phenotypes of the children are all different – blood group A, blood group B, blood group AB and blood group O.

What are the genotypes of the parents?

A $AA \times BB$

B $AO \times BO$

C $AA \times BO$

D $AO \times BB$

14. Which one of the following statements about sex-linked traits is true?

A A female transmits her sex-linked traits to her daughters only.

B A male transmits his sex-linked traits to his sons only.

C A male transmits his sex-linked traits to his grandchildren via his daughters only.

D A female transmits her sex-linked traits to her grandchildren via her sons only.

15. Colour blindness is a sex-linked trait. A man with normal vision marries a woman with normal vision. They have a son who is colour blind.

What is the chance of their next son being colour blind?

A no chance

B 1 in 2

C 1 in 3

D 1 in 4

16. The function of the seminal vesicles is to

A produce nutrients for sperm

B allow sperm to mature

C store sperm temporarily

D produce testosterone.

[Turn over

17. Which of the following is the sequence of events following fertilisation?

A Cleavage → Differentiation → Implantation

B Implantation → Differentiation → Cleavage

C Differentiation → Implantation → Cleavage

D Cleavage → Implantation → Differentiation

18. The electrocardiogram shown below records the beat of a human heart.

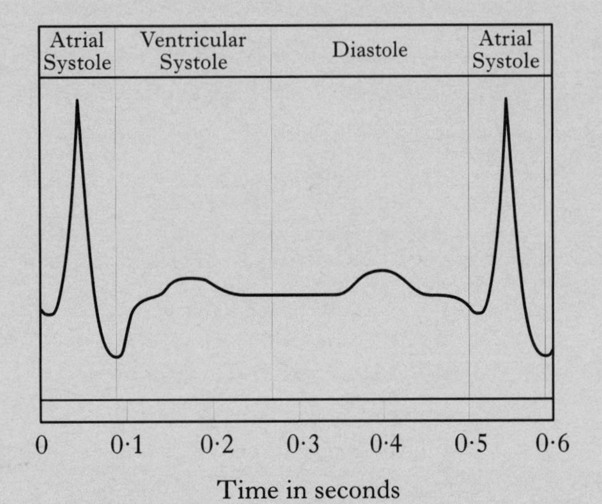

What is the heart rate?

A 60 beats/minute

B 70 beats/minute

C 75 beats/minute

D 120 beats/minute

19. Which of the following is **not** a function of the lymphatic system?

A It returns excess tissue fluid to the blood.

B It transports fat from the small intestine.

C It destroys bacteria.

D It causes the clotting of blood at wounds.

20. The following data refer to the breathing of an athlete resting and just after a race.

	Breathing Rate (per minute)	*Tidal Volume* (litres)	*% Carbon dioxide in exhaled air*
Resting	12	0·3	5
After race	24	1	5

Assuming the rate of breathing remains constant, what would be the volume of carbon dioxide breathed out during the first two minutes after exercise?

A 0·18 litres

B 0·36 litres

C 1·2 litres

D 2·4 litres

21. Which line in the table below identifies correctly the sites of secretion of the hormones ADH and glucagon?

	ADH	*Glucagon*
A	pituitary gland	liver
B	kidney	liver
C	kidney	pancreas
D	pituitary gland	pancreas

22. The graph below shows the blood glucose concentrations of two women before and after each swallowed 50 g of glucose.

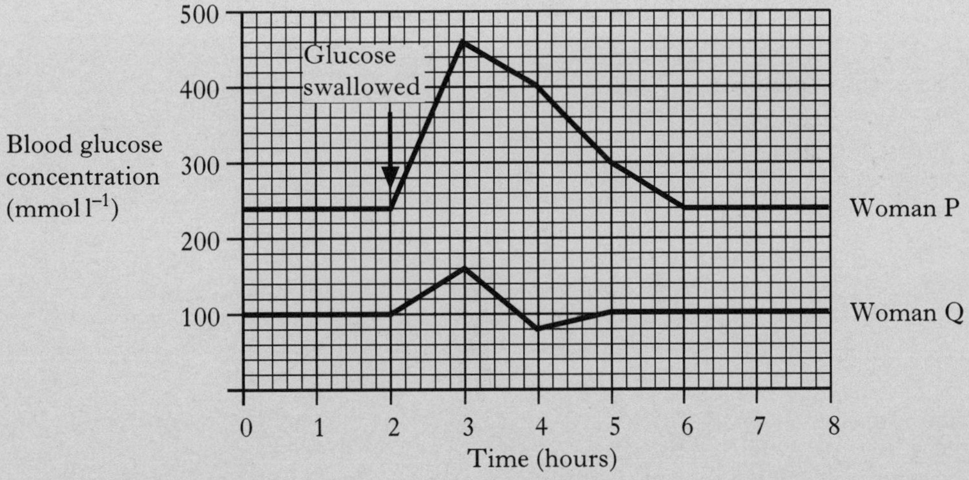

When did the rate of change of blood glucose concentration of the two women differ most?

A Between hours 2 and 3

B Between hours 3 and 4

C Between hours 4 and 5

D Between hours 5 and 6

23. Both nerves and hormones are involved in the control of homeostatic mechanisms.

In which of the following are the homeostatic mechanisms related correctly to their principal methods of control?

	Osmoregulation	*Body Temperature*
A	Nerves	Hormones
B	Hormones	Nerves
C	Nerves	Nerves
D	Hormones	Hormones

24. The human cerebrum has a highly convoluted surface. This increased surface area allows an

A increase in the types of neurones present

B increased blood supply to the brain

C increased number of interconnections between neurones

D increase in the amount of white matter on the surface.

25. Which parts of the body are controlled by the largest motor area of the cerebrum?

A Hands and lips

B Feet and hands

C Legs and feet

D Legs and arms

26. Vision in dim light is improved by the rods having

A diverging neural pathways

B converging neural pathways

C reflex neural pathways

D peripheral neural pathways.

27. The retrieval of information from long term memory is often aided by remembering the situation in which the information was encoded. This is described as using

A contextual cues

B chunking techniques

C rehearsal methods

D memory span.

28. The transformation of information into a form that memory can accept is called

A shaping

B retrieval

C encoding

D storage.

[Turn over

29. The graph shows how the algal population in a freshwater lake is affected by changes in the nitrate concentration in the water and by the changes in light intensity over the year.

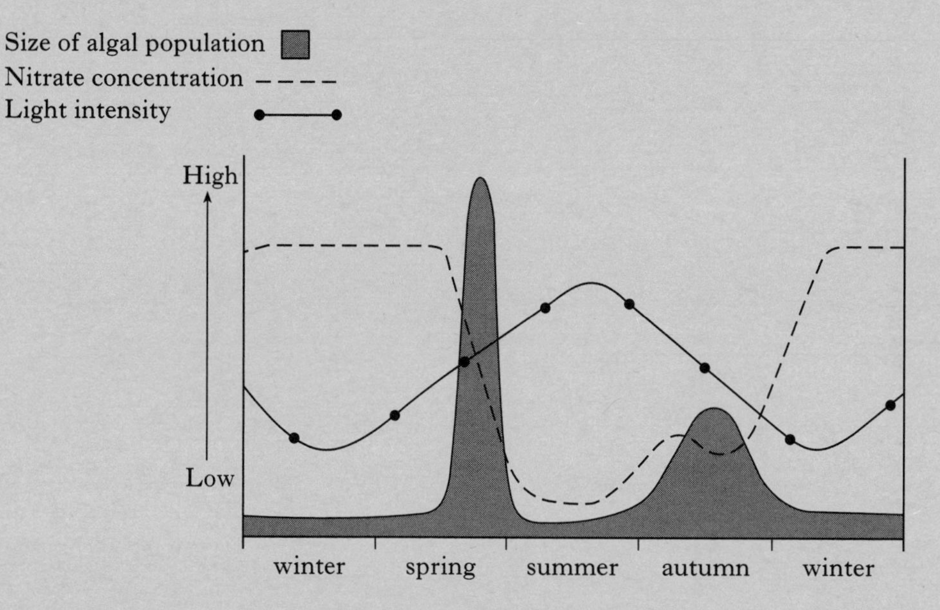

Which of the following statements may be deduced from the graph?

A The increase in the algal population in the spring is triggered by a high concentration of nitrate.

B The changes in size of the algal population are related directly to changes in light intensity.

C The increasing concentration of nitrate in the autumn is followed by an increase in the algal population.

D The size of the algal population is inversely proportional to light intensity.

30. The diagram below shows the population of a country as a percentage distribution by age.

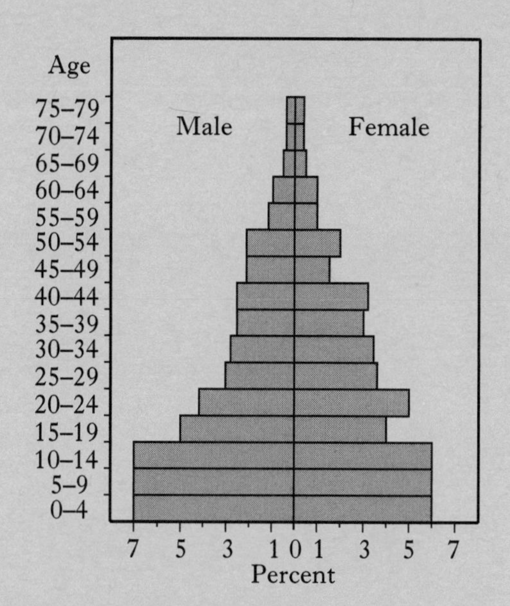

What percentage of the population is under 15 years of age?

A 21%

B 39%

C 48%

D 49%

Candidates are reminded that the answer sheet MUST be returned INSIDE this answer booklet.

[Turn over for Section B]

DO NOT WRITE IN THIS MARGIN

Marks

SECTION B

All questions in this section should be attempted.

1. The diagram shows the sequence of events as a white blood cell engulfs and destroys a bacterium.

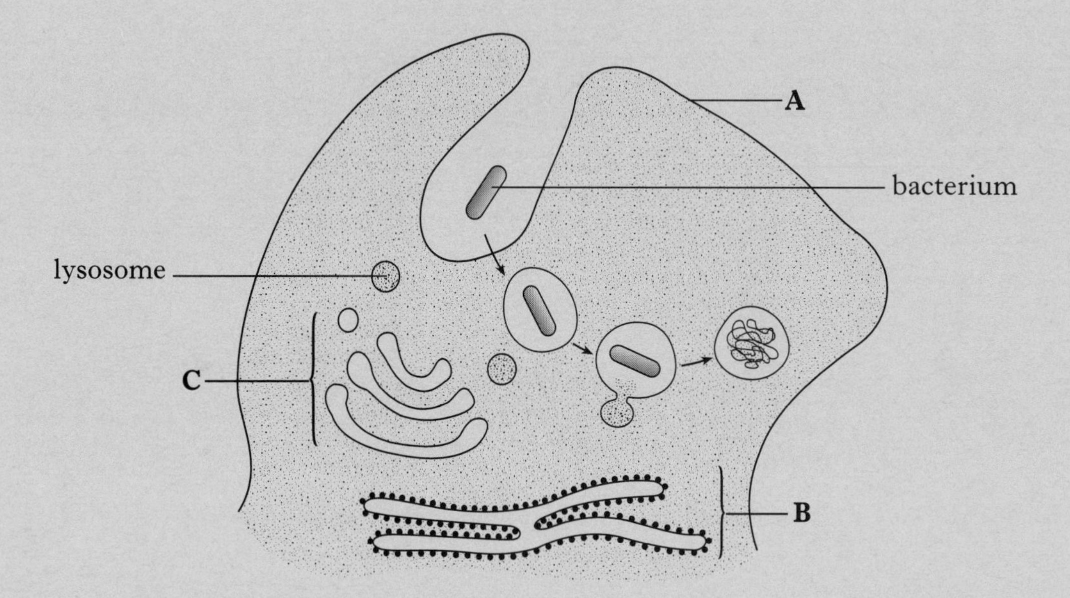

(*a*) Name the structures **A**, **B** and **C**.

A ______________________________

B ______________________________

C ______________________________ **2**

(*b*) (i) What is the name given to this process of engulfing the bacterium?

______________________________ **1**

(ii) Describe the roles of structures **B** and **C** in the cell.

B ______________________________

______________________________ **1**

C ______________________________

______________________________ **1**

(*c*) What does the lysosome contain?

______________________________ **1**

DO NOT WRITE IN THIS MARGIN

Marks

1. (continued)

(*d*) (i) Analysis of a blood sample yielded the following blood cell counts.

Cell type	*Number per mm^3* (× 1000)
White	8
Red	5600

Express as a simple ratio the number of white cells to red cells in this blood sample.

Space for calculation

______ : ______ **1**

(ii) Predict how the proportion of white cells to red cells would change if a person was suffering from influenza.

______________________________ **1**

(*e*) Explain how the shape of the red blood cell is related to its function.

____

____

____

______________________________ **2**

[Turn over

DO NOT WRITE IN THIS MARGIN

Marks

2. The graph below shows oxygen-haemoglobin dissociation curves at 37 °C and at 38 °C.

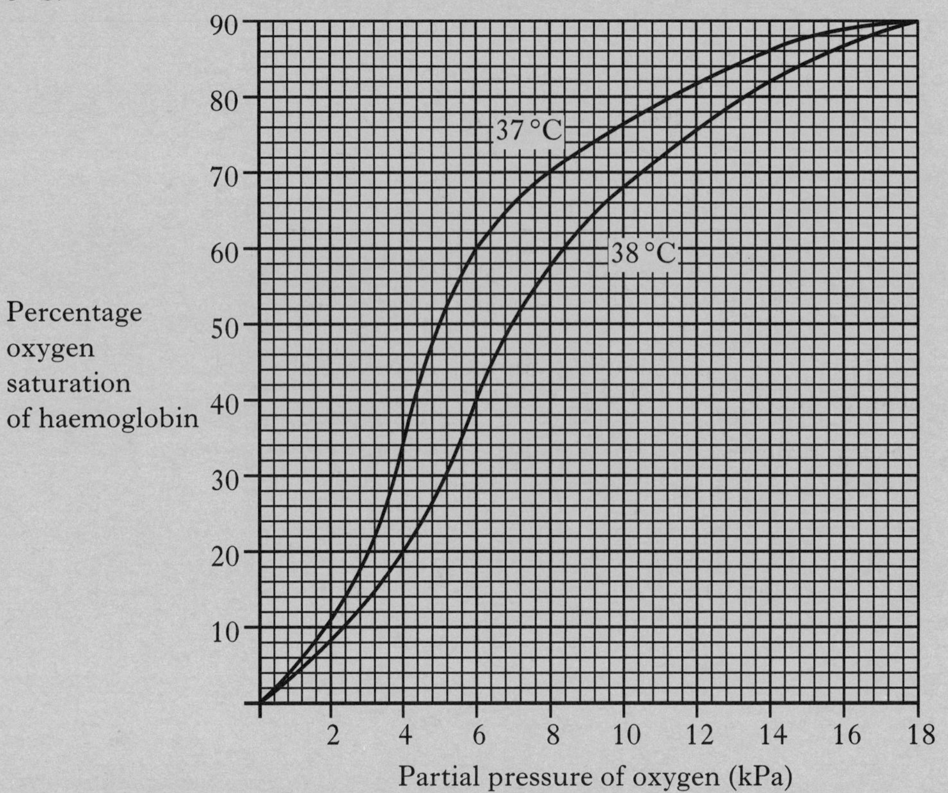

(*a*) (i) Complete the table below to show the change in percentage oxygen saturation of haemoglobin at 37 °C and 38 °C when the partial pressure drops from 18 to 6 kPa.

Partial pressure kPa	*Percentage oxygen saturation of haemoglobin*	
	37 °C	38 °C
18		
6		
change		

2

(ii) Explain why this change in percentage oxygen saturation of haemoglobin improves the efficiency of working muscles.

____

____

______________________________ **2**

(*b*) The partial pressure of oxygen in fresh air is 20 kPa. The partial pressure of oxygen in the alveoli is 16 kPa. Explain why there is a lower value for oxygen in the alveoli.

____

______________________________ **1**

DO NOT WRITE IN THIS MARGIN

Marks

3. Thalassaemia is an inherited blood disorder in which haemoglobin is affected. The condition illustrates incomplete dominance in which the recessive allele has a partial effect. Heterozygous individuals show mild symptoms.

The diagram below shows the incidence of thalassaemia in three generations of a family.

- male with mild thalassaemia
- female with mild thalassaemia
- male with severe thalassaemia
- unaffected female

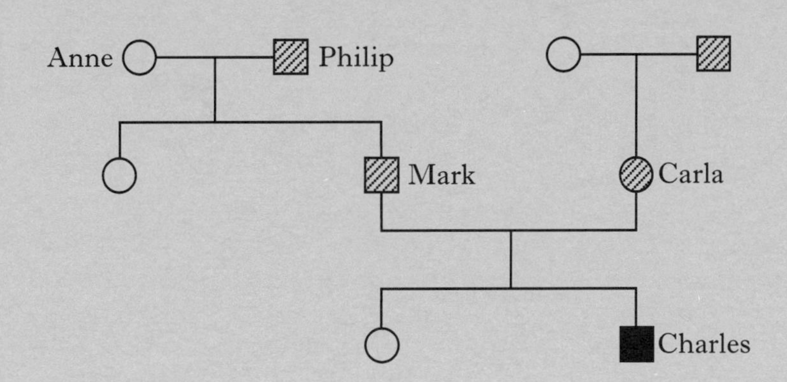

(*a*) Using the symbol Hb^A to represent the allele for normal haemoglobin, and the symbol Hb^B to represent the recessive allele, complete the table to show the genotypes of Anne, Philip and Charles.

Individual	*Genotype*
Anne	
Philip	
Charles	

2

(*b*) Mark and Carla have a third child. What is the percentage chance that the child will have the same genotype as the parents?

Space for calculation

__________ % **1**

(*c*) Haemoglobin is found in red blood cells. Where in the body are red cells manufactured and destroyed?

Manufactured ____________________ Destroyed ____________________ **2**

[Turn over

DO NOT WRITE IN THIS MARGIN

Marks

4. The diagram shows part of an enzyme molecule.

P represents a molecule in the chain which forms the primary structure of the protein.

Q is a bond which links these molecules.

R is a bond which maintains the secondary structure of the protein.

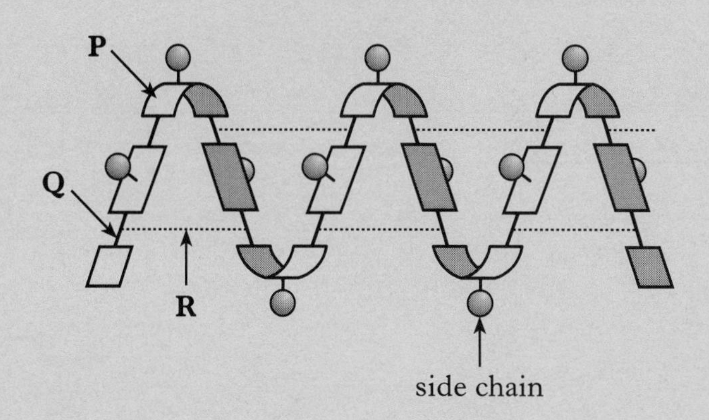

(*a*) (i) Identify molecule **P**.

____________________ **1**

(ii) Name bonds **Q** and **R**.

Q ____________________ **R** ____________________ **2**

(*b*) (i) Why are some digestive enzymes produced in an inactive form?

____

____ **1**

(ii) Give an example of a substance which can act as an enzyme activator.

____ **1**

(*c*) The table below contains information about the effects of environmental factors on enzymes. Complete the table.

Factor	*Type of change*	*Effect on enzyme structure*
high temperature		alters active site
	mutation	

3

DO NOT WRITE IN THIS MARGIN

Marks

5. The diagram below represents two stages in the chemistry of respiration in a cell which is respiring aerobically.

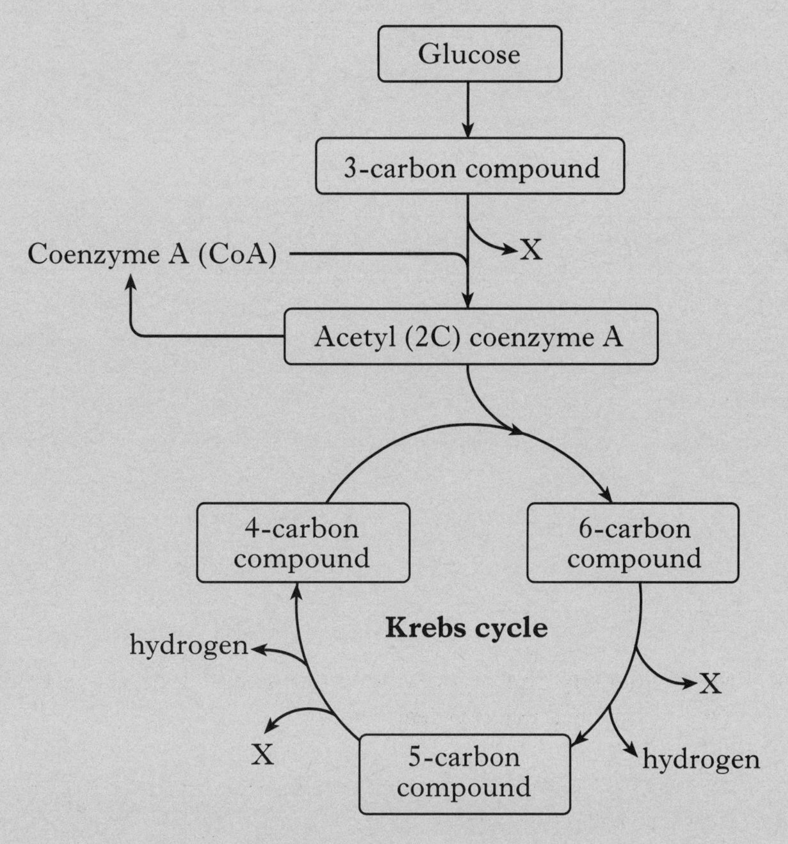

(*a*) Name the 3-carbon compound and the 6-carbon compound.

3C ______________________

6C ______________________ **2**

(*b*) Draw a line across the diagram to indicate where this series of reactions would stop if oxygen were not available. **1**

(*c*) State the precise location of the Krebs cycle within the cell.

____ **1**

(*d*) Complete the table below to name product **X** and to describe what happens to each of the products.

Product	*Fate of product*
X ____________	
Hydrogen	

2

DO NOT WRITE IN THIS MARGIN

Marks

6. The diagram below shows the liver with its blood supply and an associated organ.

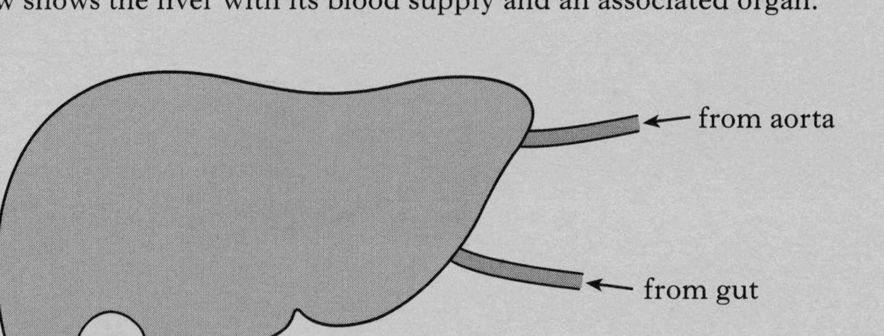

(*a*) Name the liquid stored in organ **X**.

______________________________ **1**

(*b*) Complete the table to identify the blood vessels carrying blood to and from the liver and the type of blood carried by each vessel.

Blood supply	*Name of blood vessel*	*Deoxygenated or oxygenated blood*
from aorta		
from gut		
to vena cava		

3

(*c*) Complete the following sentences by underlining one option in each set of brackets.

Deamination occurs in the {gall bladder / kidney / liver}. During this process excess {nucleic acids / amino acids / fatty acids} are broken down. A waste product of this process is {urea / glycogen / glucose} which is carried in the blood to the {urinary bladder / kidney / liver} where it is removed from the blood. **2**

(*d*) Name a hormone involved in the processing of carbohydrate in the liver.

______________________________ **1**

DO NOT WRITE IN THIS MARGIN

Marks

7. An investigation was carried out to determine the urea content of three samples taken from a healthy individual. The three fluids sampled were:

A plasma from the renal artery

B plasma from the renal vein

C urine from the urethra.

One urease tablet was added to each sample as shown in the diagram below.

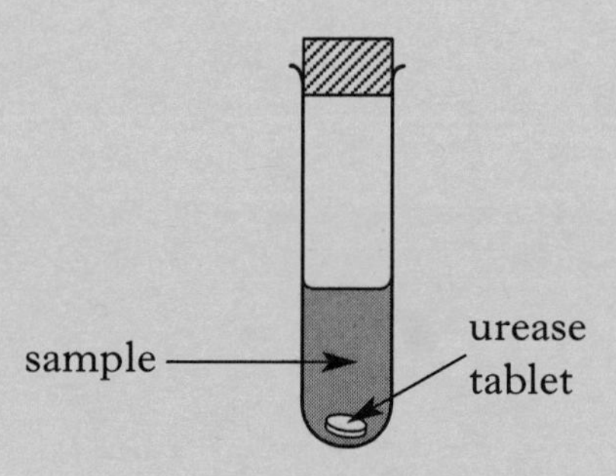

The urease catalyses the breakdown of urea to ammonium carbonate.

$$\textbf{urea + water} \xrightarrow{\textit{urease}} \textbf{ammonium carbonate}$$

Once the reaction had finished, the samples were analysed to determine the concentration of ammonium carbonate.

(*a*) List **two** variables which would have to be kept constant for a valid comparison of the three samples.

1 ______________________________

2 ______________________________ **2**

(*b*) The table below shows the results of this investigation.

Complete the table to identify the three samples.

Fluid sample (A, B or C)	*Ammonium carbonate concentration* (g/litre)
	0·16
	16·7
	0·52

1

(*c*) Suggest how the investigation could be improved to ensure the reliability of the results.

____

______________________________ **1**

[Turn over

DO NOT WRITE IN THIS MARGIN

Marks

8. The diagram below shows stages in the control of a person's body temperature.

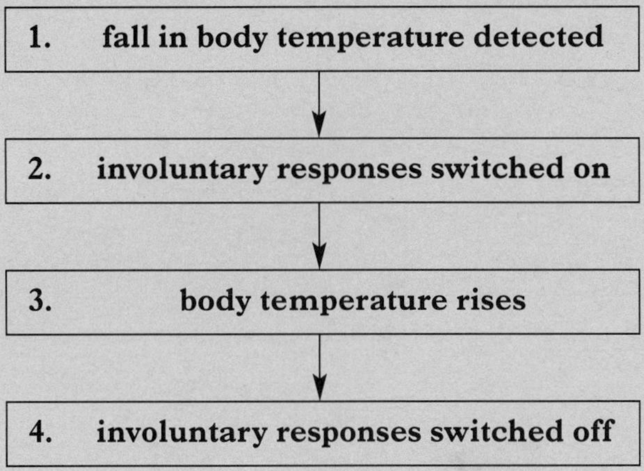

1. fall in body temperature detected

↓

2. involuntary responses switched on

↓

3. body temperature rises

↓

4. involuntary responses switched off

(*a*) Where in the body is the temperature monitoring centre located?

______________________________ **1**

(*b*) State **two** involuntary responses which may be switched on in this individual.

1 ______________________________

2 ______________________________ **2**

(*c*) Explain the role of negative feedback control in this process.

____

____

______________________________ **2**

(*d*) The table below shows the surface area and mass of a baby and an adult.

	Surface area (m^2)	*Mass* (kg)
Baby	0·2	4
Adult	2·0	80

With reference to the table, explain why babies are more susceptible to hypothermia than adults.

____

____

____

______________________________ **2**

DO NOT WRITE IN THIS MARGIN

Marks

9. The graph shows increases in brain volume at four stages of human evolution over the last four million years.

The bars indicate the range of volumes and the mid (median) volume.

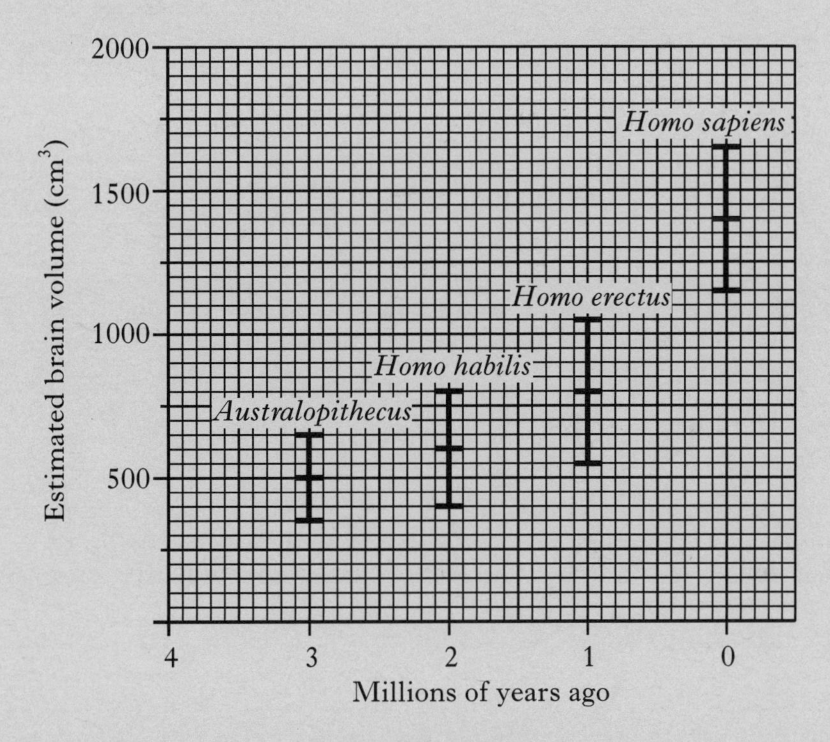

(*a*) State the range of brain volume for *Homo habilis*.

____________ to ____________ cm^3 **1**

(*b*) Complete the table below for *Homo sapiens*.

Species	*Median volume* (cm^3)	*Percentage increase*
Australopithecus	500	–
Homo habilis	600	20%
Homo erectus	800	33%
Homo sapiens		

2

(*c*) Name the part of the brain which

(i) contributes most to brain volume;

________________________ **1**

(ii) links the two hemispheres;

________________________ **1**

(iii) is rich in the receptor NMDA.

________________________ **1**

DO NOT WRITE IN THIS MARGIN

Marks

10. The diagram represents a motor neurone.

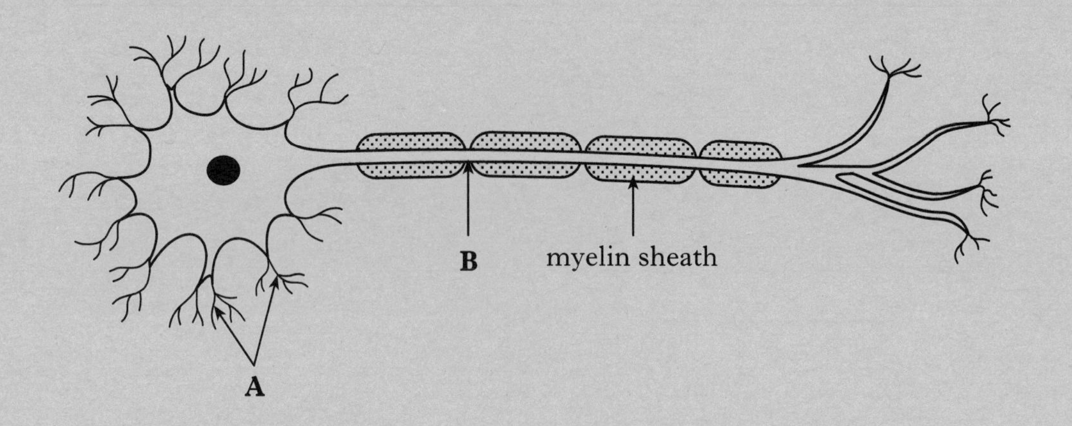

(*a*) Name the nerve fibres **A** and **B**.

A ______________________ **B** ______________________ **1**

(*b*) The table below describes features of somatic and autonomic motor neurone function.

Complete the table.

Feature	*Somatic*	*Autonomic*
type of control (conscious/unconscious)		
example of target muscle		uterine muscle
example of neurotransmitter		noradrenaline

2

(*c*) State the effect of sympathetic stimulation on the:

1 heart rate ______________________

2 digestive system ______________________

3 skin arterioles ______________________ **3**

(*d*) The sympathetic and parasympathetic nervous systems often influence organs in opposite ways. What term describes this opposing effect?

______________________ **1**

DO NOT WRITE IN THIS MARGIN

Marks

11. The diagram shows a simplified outline of the nitrogen cycle.

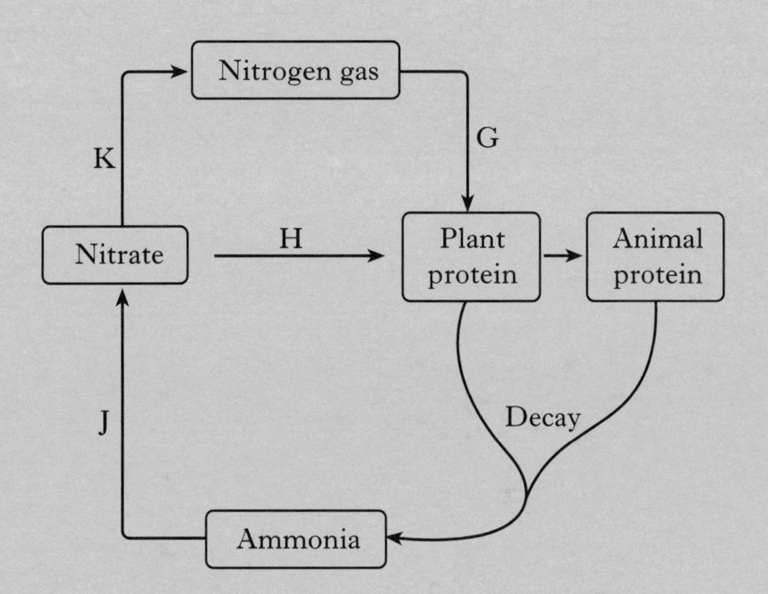

(*a*) The table below shows some of the processes involved in the nitrogen cycle.

Complete the table using information from the diagram.

Label	*Type of bacteria*	*Process in nitrogen cycle*
G		trap atmospheric nitrogen
	nitrifying	
K		convert nitrate to nitrogen gas

2

(*b*) Nitrate is lost from the soil by leaching.

Describe the effect of nitrate pollution on fresh water environments.

____

____

____

____

____

____

____ **3**

[Turn over

DO NOT WRITE IN THIS MARGIN

Marks

12. **Graphs 1** and **2** below contain information about measles in Scotland from 1974 to 1999.

Graph 1 – Cases of measles

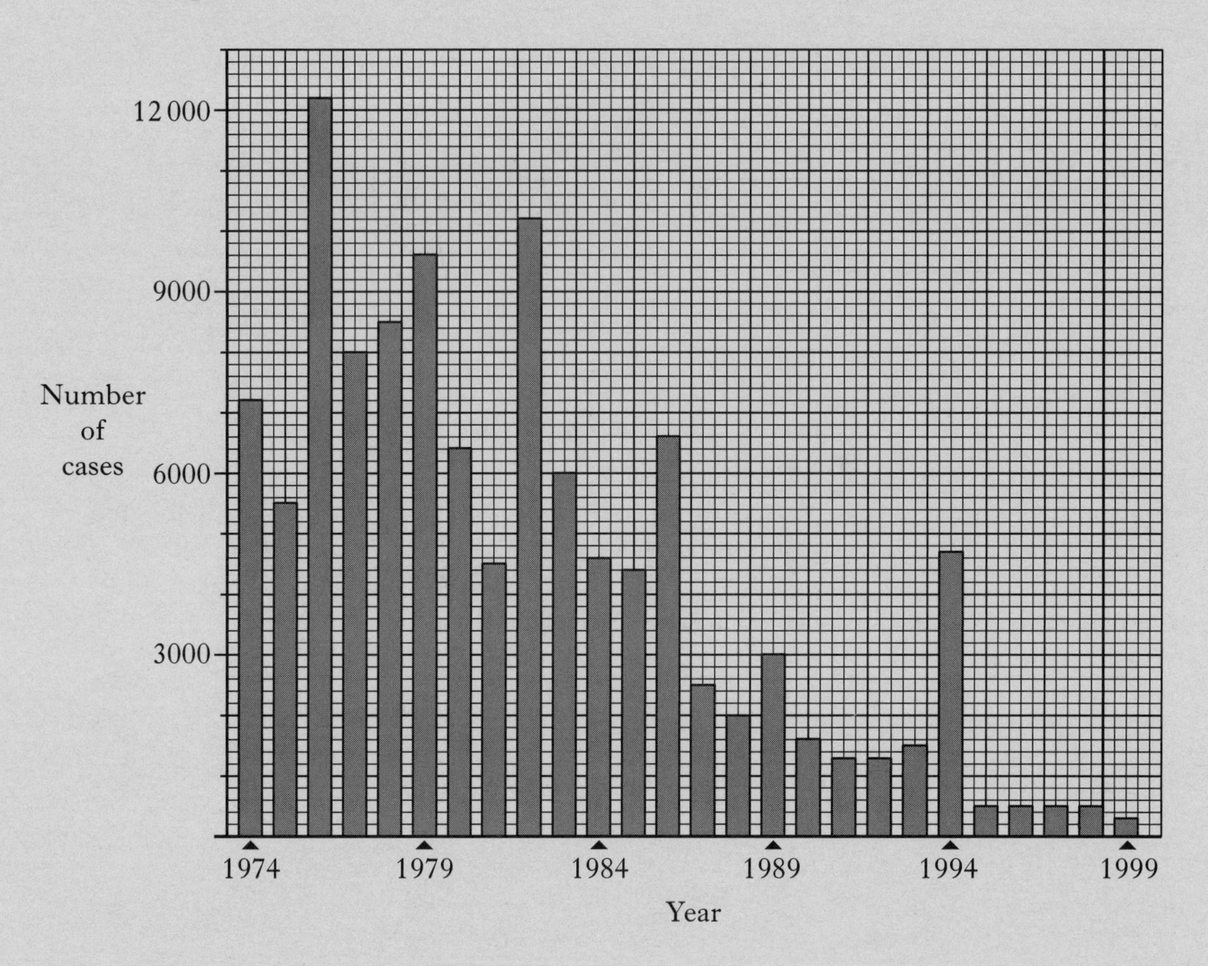

Graph 2 – Deaths from measles

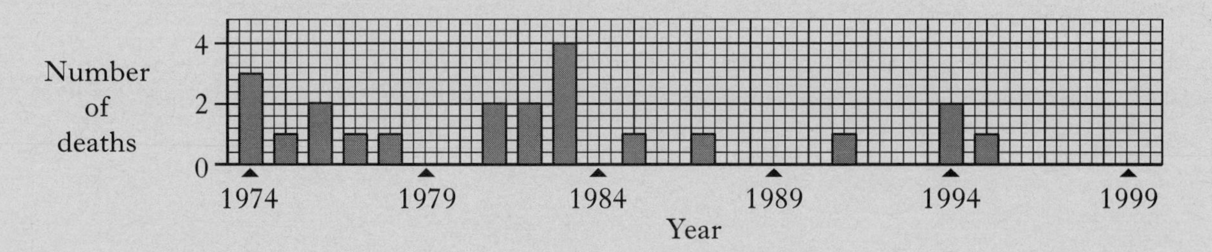

(*a*) From **Graph 1** construct a table to show the years in which measles cases were at their lowest and highest levels. The table should include the number of cases in each year. **2**

DO NOT WRITE IN THIS MARGIN

Marks

12. (continued)

(*b*) Suggest a reason for the trend in measles cases shown in **Graph 1**.

____

____ **1**

(*c*) From **Graphs 1** and **2**, what percentage of individuals who contracted measles in 1995 died from the disease?

Space for calculation

____________% **1**

[Turn over

13. The following information relates to the impact of human activities on the carbon cycle in the year 2000.

Table 1 World carbon reserves

Reservoir	*Mass of carbon stored* (billions of tons of carbon per year)
Oceans	35 000
Fossil fuels	10 000
Soil	1500
Atmosphere	500
Plants	500

Table 2 Mass of carbon released by human activity

Activity	*Mass of carbon released* (billions of tons of carbon per year)
Burning of fossil fuels	5·5
Deforestation	1·5

Table 3 Annual carbon gain by the atmosphere and the oceans

Reservoir	*Mass of carbon gained* (billions of tons of carbon per year)
Atmospheric carbon	3·3
Oceanic carbon	2·0

DO NOT WRITE IN THIS MARGIN

Marks

13. (continued)

(*a*) Construct a bar chart to illustrate the data in **Table 1**.

(Additional graph paper, if required, can be found on page 28.)

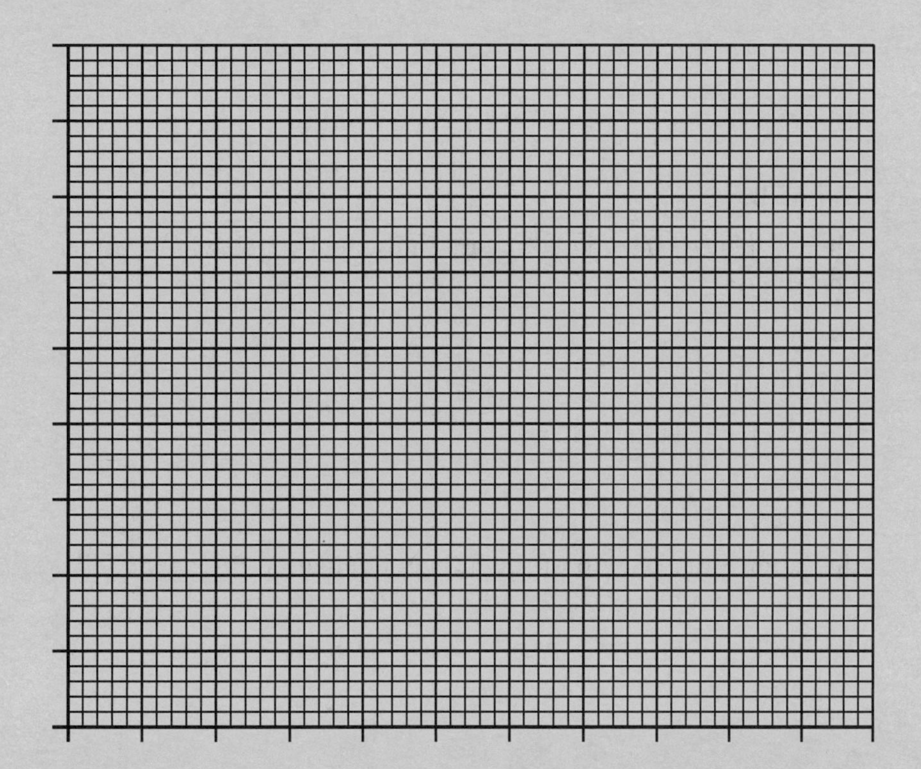

2

(*b*) It is estimated that only 10% of fossil fuel reserves are available for human use. Using information from **Tables 1** and **2** calculate how long these reserves will last.

Space for calculation

________ 1

(*c*) Use the information from **Tables 1** and **3** to estimate the number of years it will take for atmospheric carbon to exceed 550 billion tons.

Space for calculation

________ 1

(*d*) State **two** likely consequences of increased atmospheric carbon levels on global climate patterns.

1 ________________________

2 ________________________ 2

DO NOT WRITE IN THIS MARGIN

Marks

SECTION C

Both questions in this section should be attempted.

Note that each question contains a choice.

Questions 1 and 2 should be attempted on the blank pages which follow.

Supplementary sheets, if required, may be obtained from the invigilator.

Labelled diagrams may be used where appropriate.

1. Answer **either** A **or** B.

A. Discuss the influence of others on an individual's behaviour under the following headings:

(i) Social facilitation; **3**

(ii) Deindividuation; **3**

(iii) Influences that change beliefs. **4**

(10)

OR

B. Discuss the exponential growth of the human population under the following headings:

(i) Demographic trends; **2**

(ii) Agriculture; **4**

(iii) Disease. **4**

(10)

In question 2 ONE mark is available for coherence and ONE mark is available for relevance.

2. Answer **either** A **or** B.

A. Describe the influence of hormones on the testes. **(10)**

OR

B. Describe the events which take place in the first half of the menstrual cycle. **(10)**

[*END OF QUESTION PAPER*]

DO NOT WRITE IN THIS MARGIN

SPACE FOR ANSWERS

DO NOT WRITE IN THIS MARGIN

SPACE FOR ANSWERS

ADDITIONAL GRAPH PAPER FOR QUESTION 13(*a*)

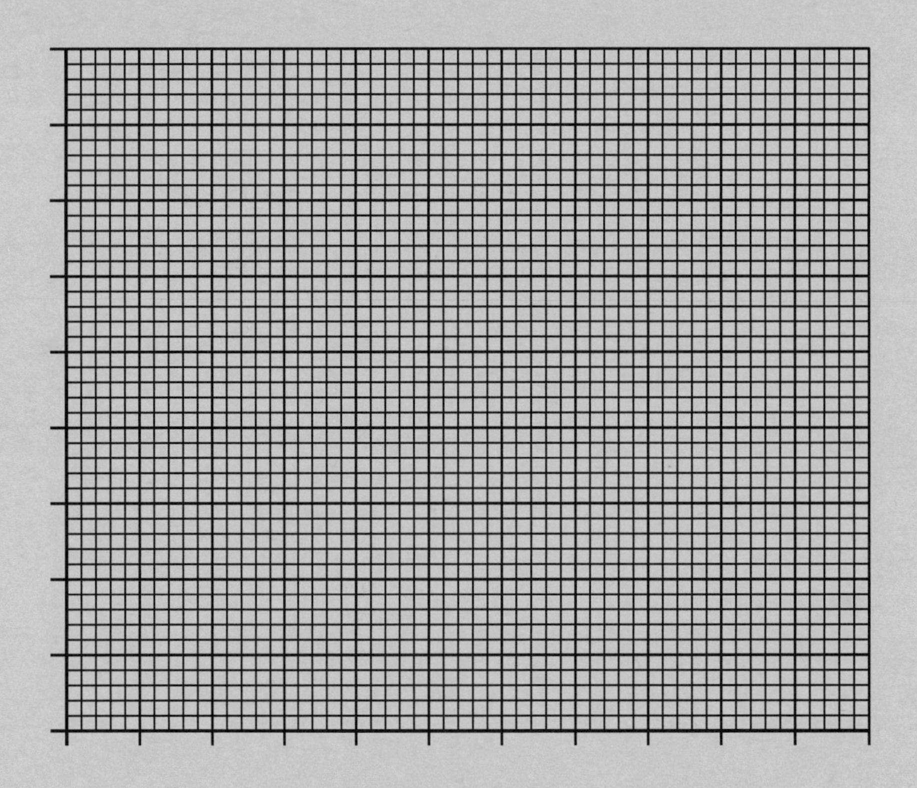

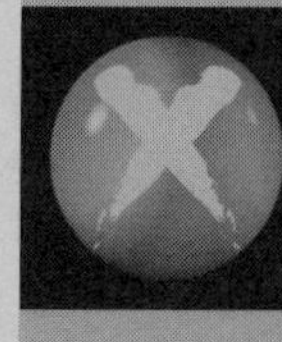

[BLANK PAGE]

FOR OFFICIAL USE

Total for Sections B & C

X009/301

NATIONAL QUALIFICATIONS 2004

WEDNESDAY, 19 MAY 1.00 PM – 3.30 PM

HUMAN BIOLOGY HIGHER

Fill in these boxes and read what is printed below.

Full name of centre

Town

Forename(s)

Surname

Date of birth
Day Month Year

Scottish candidate number

Number of seat

SECTION A—Questions 1–30

Instructions for completion of Section A are given on page two.

SECTIONS B AND C

1 (a) All questions should be attempted.

(b) It should be noted that in **Section C** questions 1 and 2 each contain a choice.

2 The questions may be answered in any order but all answers are to be written in the spaces provided in this answer book, and must be written clearly and legibly in ink.

3 Additional space for answers and rough work will be found at the end of the book. If further space is required, supplementary sheets may be obtained from the invigilator and should be inserted inside the **front** cover of this book.

4 The numbers of questions must be clearly inserted with any answers written in the additional space.

5 Rough work, if any should be necessary, should be written in this book and then scored through when the fair copy has been written.

6 Before leaving the examination room you must give this book to the invigilator. If you do not, you may lose all the marks for this paper.

SCOTTISH QUALIFICATIONS AUTHORITY

SECTION A

Read carefully

1 Check that the answer sheet provided is for Human Biology Higher (Section A).

2 Fill in the details required on the answer sheet.

3 In this section a question is answered by indicating the choice A, B, C or D by a stroke made in **ink** in the appropriate place in the answer sheet—see the sample question below.

4 For each question there is only **one** correct answer.

5 Rough working, if required, should be done only on this question paper—or on the rough working sheet provided—**not** on the answer sheet.

6 At the end of the examination the answer sheet for Section A **must** be placed **inside** this answer book.

Sample Question

The digestive enzyme pepsin is most active in the

A mouth

B stomach

C duodenum

D pancreas.

The correct answer is **B**—stomach. A **heavy** vertical line should be drawn joining the two dots in the appropriate box in the column headed **B** as shown in the example on the answer sheet.

If, after you have recorded your answer, you decide that you have made an error and wish to make a change, you should cancel the original answer and put a vertical stroke in the box you now consider to be correct. Thus, if you want to change an answer D to an answer B, your answer sheet would look like this:

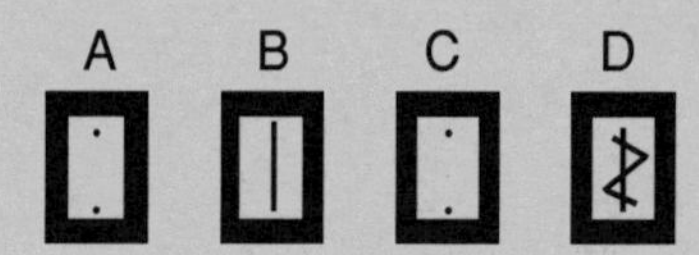

If you want to change back to an answer which has already been scored out, you should enter a tick (✓) to the **right** of the box of your choice, thus:

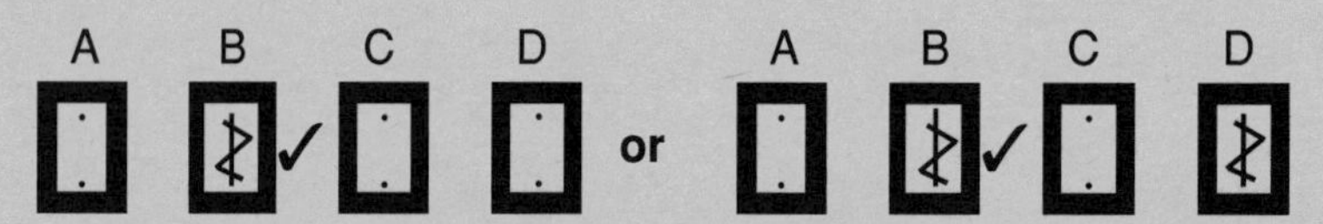

SECTION A

All questions in this section should be attempted.

Answers should be given on the separate answer sheet provided.

1. In respiration, the sequence of reactions resulting in the conversion of glucose to pyruvic acid is called

A the Krebs cycle

B the citric acid cycle

C glycolysis

D the cytochrome chain.

2. Which of the following is an insoluble polysaccharide?

A Glycogen

B Glucose

C Sucrose

D Maltose

3. Which of the following is **not** a function of lipids?

A Nerve insulation

B Vitamin transport

C Energy storage

D Oxygen transport

4. Which of the following processes requires infolding of the cell membrane?

A Diffusion

B Phagocytosis

C Active transport

D Osmosis

5. The formation of new viruses involves the following stages:

X viral protein coats are synthesised

Y host cell metabolism is taken over by virus

Z viral nucleic acid is replicated.

The correct order in which these stages occur is

A X → Z → Y

B Y → X → Z

C Z → X → Y

D Y → Z → X.

6. A sex-linked gene carried on the X-chromosome of a man will be transmitted to

A 50% of his male children

B 50% of his female children

C 100% of his male children

D 100% of his female children.

7. The family tree shows the inheritance of red-green colour blindness in humans. Red-green colour blindness is a recessive, sex-linked condition.

○ unaffected female

● affected female

□ unaffected male

■ affected male

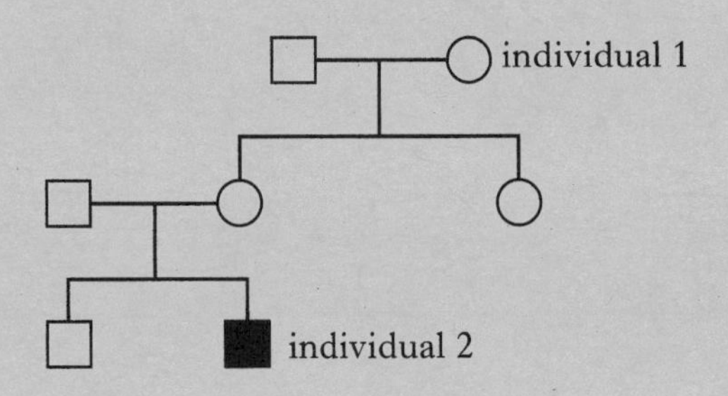

Which line in the table describes correctly the genotypes of individual 1 and individual 2?

	Individual 1	*Individual 2*
A	X^RX^R	X^RY
B	X^RX^r	X^RY
C	X^rX^r	X^rY
D	X^RX^r	X^rY

[Turn over

8. Which of the following describes the term non-disjunction?

A The failure of chromosomes to separate at meiosis

B The independent assortment of chromosomes at meiosis

C The exchange of genetic information at chiasmata

D An error in the replication of DNA before cell division

9. The diagram below shows a cross-section of a testis.

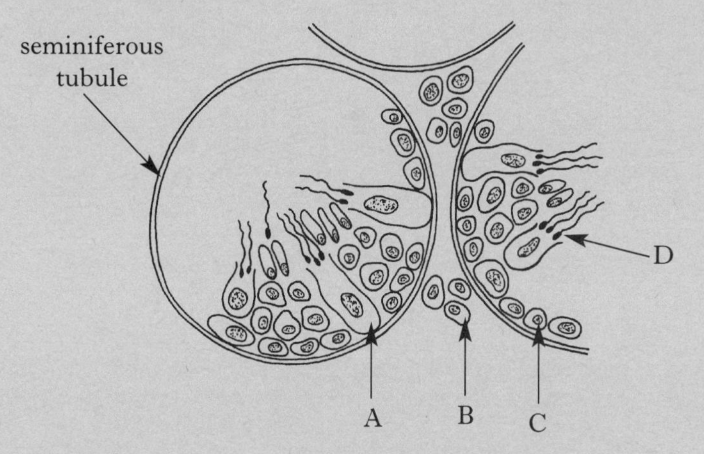

Which cell can produce testosterone?

10. The graph below shows the growth, in length, of a human fetus before birth.

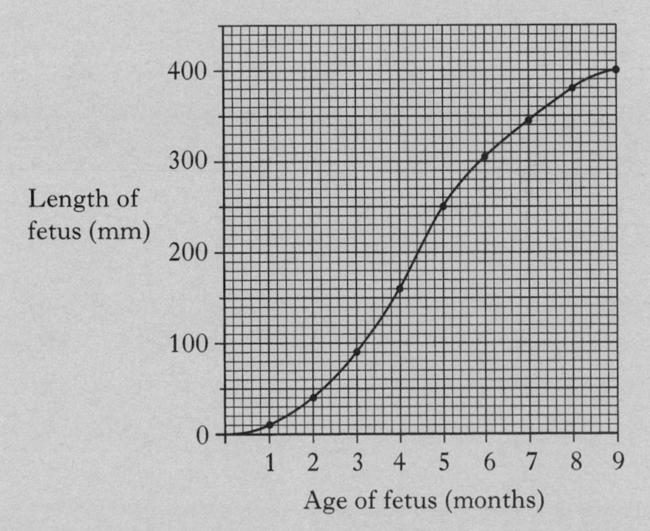

What is the percentage increase in length of the fetus during the final 4 months of pregnancy?

A 33·3

B 60·0

C 62·5

D 150·0

11. The graph below shows the dissociation curves for fetal and maternal haemoglobin.

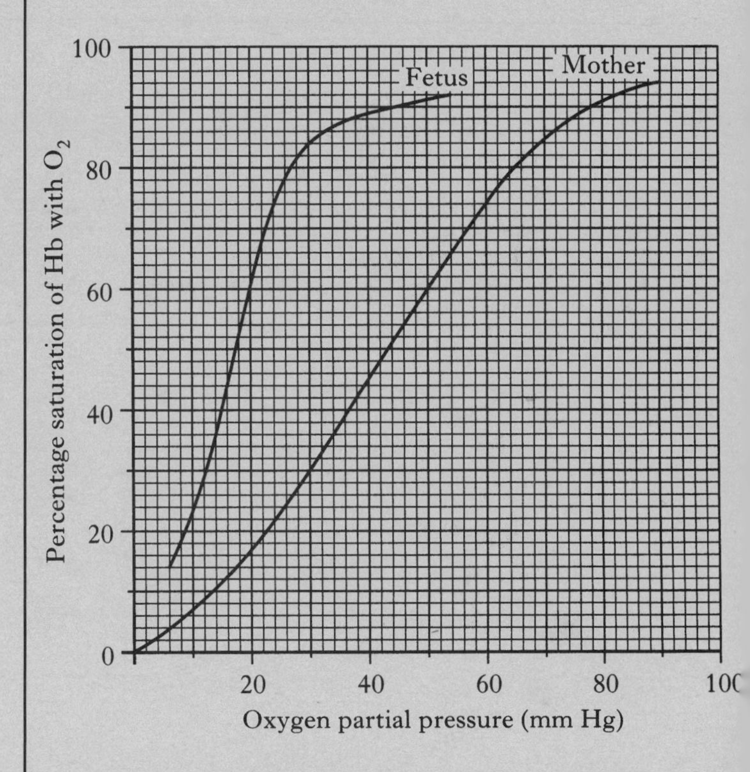

What is the difference in percentage saturation of haemoglobin between the mother and the fetus at a partial pressure of 30 mm Hg?

A 18

B 19

C 52

D 54

12. Which of the following are required for red blood cell production?

A Iron and vitamin D

B Calcium and vitamin B_{12}

C Iron and vitamin B_{12}

D Calcium and vitamin D

13. Colostrum provides a baby with

A antibodies

B antigens

C phagocytes

D lymphocytes.

14. The graph shows changes in lung volume during a breathing exercise.

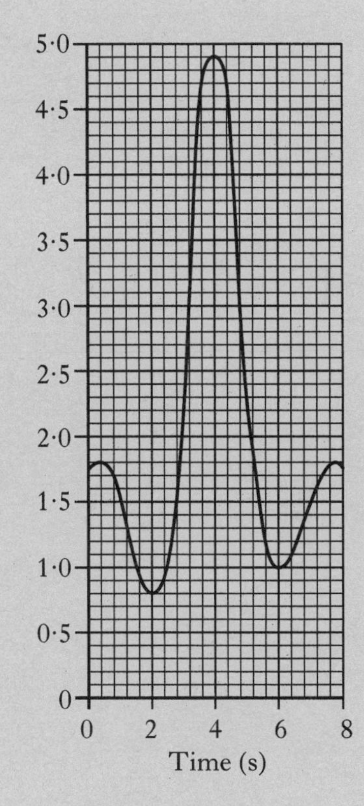

What is the volume of air inhaled between 2 and 4 seconds?

A 0·8 litres

B 3·9 litres

C 4·1 litres

D 4·9 litres

15. Which **two** blood vessels are involved in the transport of blood to and from the head?

A Carotid artery and jugular vein

B Renal artery and pulmonary vein

C Aorta and renal vein

D Hepatic artery and jugular vein

16. The table below shows the relative concentrations of certain substances in blood vessels leading to and from the liver.

(+++ = high, ++ = moderate, + = low)

Blood vessel	*Oxygen*	*Carbon dioxide*	*Urea*	*Amino acids*
1	+++	+	+	+
2	+	+++	+	+++
3	+	+++	+++	+

Which line of the table below identifies correctly the blood vessels?

	Hepatic vein	*Hepatic portal vein*	*Hepatic artery*
A	1	2	3
B	2	3	1
C	3	2	1
D	3	1	2

17. Which line of the table identifies correctly the hormones which stimulate the inter-conversion of glucose and glycogen?

	glucose → glycogen	*glycogen → glucose*
A	insulin	glucagon and adrenaline
B	glucagon and insulin	adrenaline
C	adrenaline and glucagon	insulin
D	adrenaline	glucagon and insulin

[Turn over

18. Which of the following shows the substance from which urea is produced and the site of urea production?

	Substance	*Site of production*
A	amino acid	liver
B	amino acid	kidney
C	glycogen	liver
D	glycogen	kidney

19. What is the function of the glomerulus in the production of urine?

A Collection of filtrate

B Filtration of blood

C Reabsorption of glucose

D Osmoregulation

20. The concentration of urea in plasma and urine is given in the table below.

	Plasma	*Urine*
Urea (g/100 cm^3)	0·3	1·29

By how many times has the urea been concentrated by the activity of the kidney?

A 0·23 times

B 0·39 times

C 4·3 times

D 43 times

21. The diagram below shows a body shape made up of cubes.

The surface area to volume ratio of this body is

A 4 : 1

B 6 : 1

C 15 : 4

D 29 : 8

22. The temperature monitoring centre of the brain is in the

A medulla oblongata

B cerebellum

C pituitary gland

D hypothalamus.

23. The diagram below shows the body's response to temperature change.

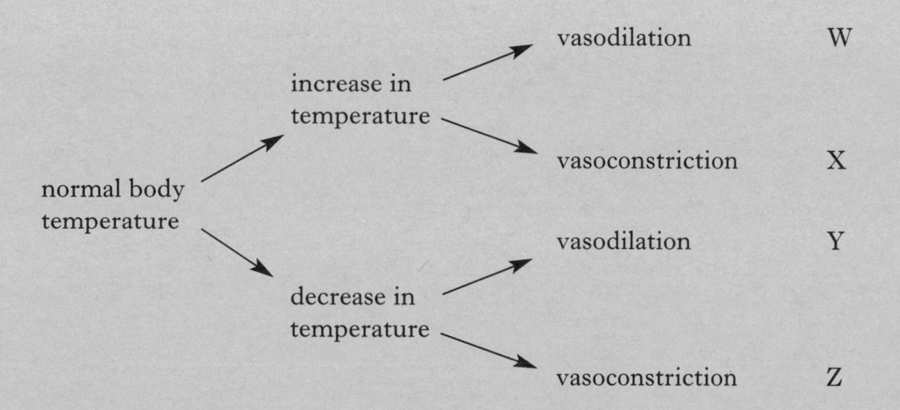

Which letters indicate negative feedback control of body temperature?

A W and Y

B W and Z

C X and Y

D X and Z

24. The peripheral nervous system contains the

A brain and spinal cord

B brain and somatic system

C spinal cord and autonomic system

D somatic system and autonomic system.

25. An investigation was carried out to determine how long it takes a student to learn the pathway through a finger maze. The student was allowed to complete the maze ten times. Which of the following pairs of factors would have to be kept the same each time?

A The time taken to complete the maze and the shape of the maze

B The number of errors made and the finger used

C The finger used and the shape of the maze

D The time taken to complete the maze and the finger used

26. Which of the following best describes social facilitation?

A Improved performance in the presence of others

B Deindividuation in competitive situations

C Discrimination behaviour shown by groups of individuals

D Shaping behaviour as seen in trial and error learning

27. Why do humans have a long period of dependency?

A To allow for learning and the development of language

B To allow bonding to take place between mother and child

C To allow for the learning of motor and sensory skills

D To allow for the growth of the brain and other major body organs

28. The diagram below shows the carbon cycle.

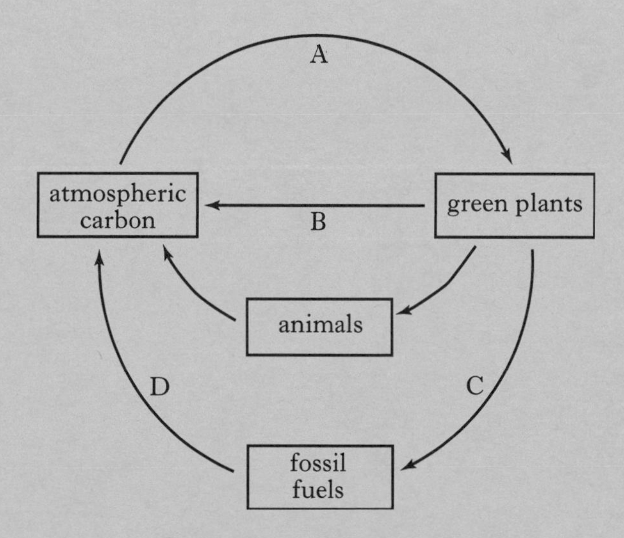

Which letter represents respiration?

[Turn over

29. In the nitrogen cycle, which of the following processes is carried out by nitrifying bacteria?

The conversion of

A nitrate to ammonia

B ammonia to nitrate

C nitrogen gas to ammonia

D nitrogen gas to nitrate.

30. An algal bloom in a loch may result from

A lack of oxygen

B lack of sunlight

C excess phosphates

D excess herbicide.

Candidates are reminded that the answer sheet MUST be returned INSIDE the front cover of this answer booklet.

DO NOT WRITE IN THIS MARGIN

Marks

SECTION B

All questions in this section should be attempted.

1. The diagram shows the role of ATP in cell metabolism.

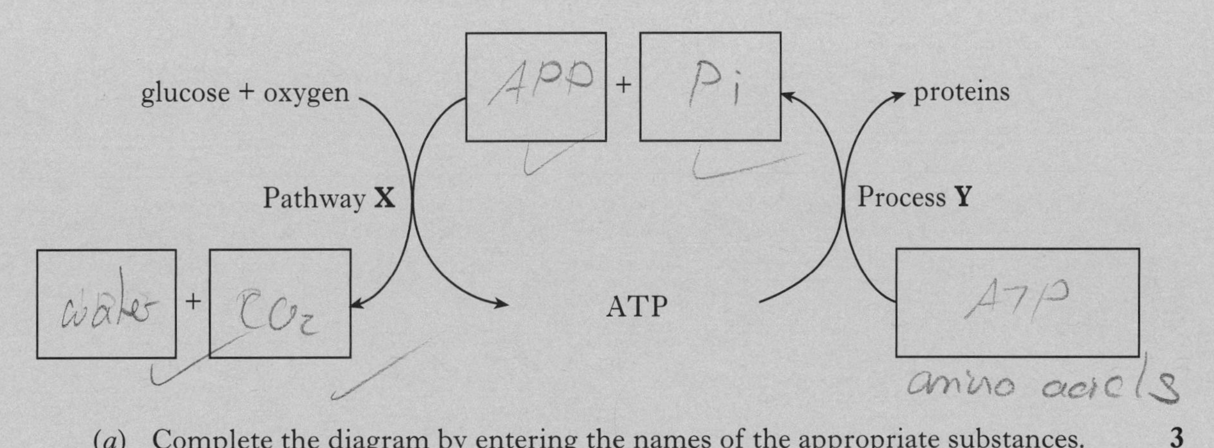

(*a*) Complete the diagram by entering the names of the appropriate substances. **3**

(*b*) (i) Name **one** stage of pathway **X** and state where it occurs in the cell.

Stage ~~respiration~~ glygolysis Location cytoplasm **1**

(ii) Name the organelle where process **Y** occurs.

ribosome **1**

(*c*) Describe **two** ways in which the diagram would be different under anaerobic conditions.

1 oxygen would not be present

2 Only 2 ATP would be produced in pathway X, Lactic acid instead of CO_2 + H_2O **2**

(*d*) Name a respiratory substrate other than glucose.

protein **1**

[Turn over

DO NOT WRITE IN THIS MARGIN

Marks

2. Sickle-cell anaemia is a blood disorder in which haemoglobin is malformed.

The diagram below shows the effect of this disorder on a red blood cell.

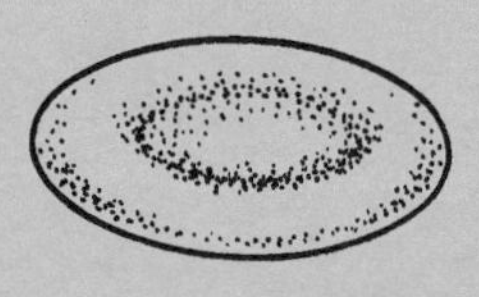
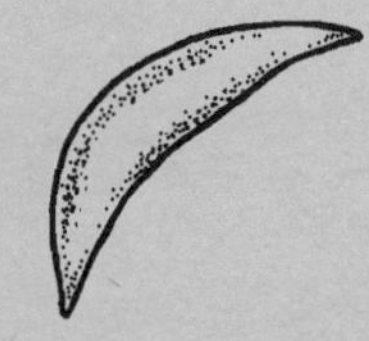
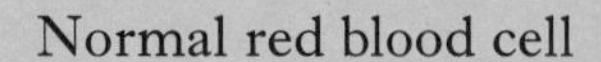

Normal red blood cell — Sickled red blood cell

The condition is not sex-linked. The allele for normal haemoglobin (**H**) is incompletely dominant to the sickle-cell allele (**h**).

Heterozygous individuals are mildly affected, whereas those with genotype **hh** are severely affected.

Two mildly affected parents have two children who are mildly affected like their parents. The parents are expecting a third child.

(*a*) Complete the Punnett square to show the possible genotypes of this child. **2**

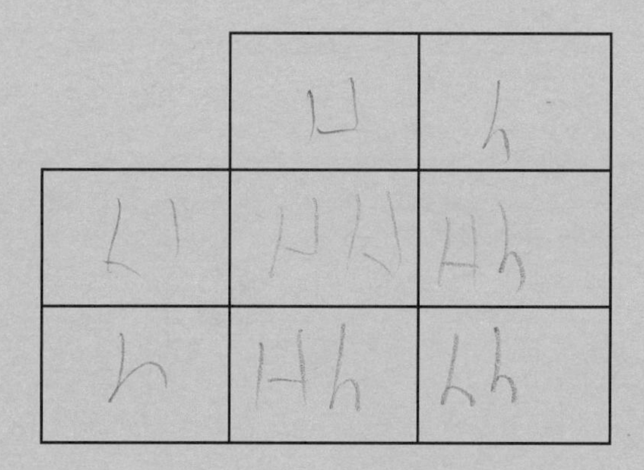

(*b*) From the Punnett square calculate the percentage chance of the child being

1 unaffected 25

2 mildly affected 50

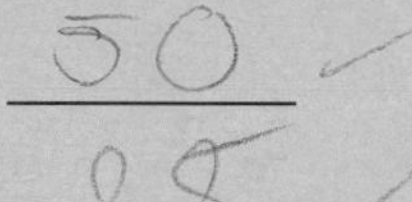

3 severely affected. 25 **1**

DO NOT WRITE IN THIS MARGIN

Marks

3. The diagram below shows a section through a lymph node.

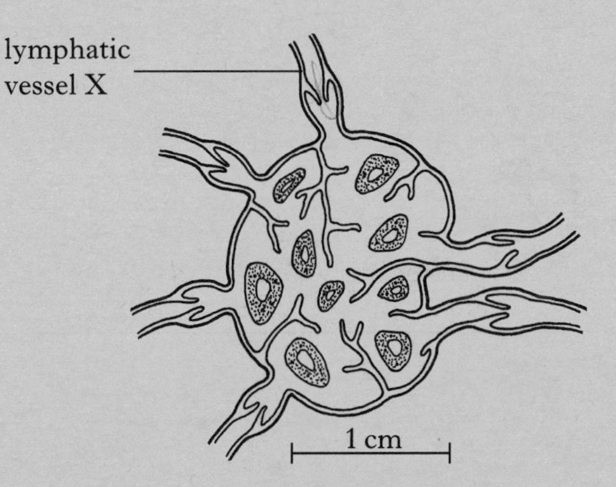

(*a*) Complete the table to name the cells found in the node, and to describe their functions.

Type of cell	*Secretion of antibodies (yes/no)*	*Type of response*
B-lymphocyte	yes ✓	humoral ✓
T-lymphocytes	no ✓	cell-mediated response
macrophage		non-specific response

3

(*b*) Add an arrow to the diagram to indicate the direction of flow of lymph in vessel X. Give a reason for your choice.

Reason because the valve is allowing lymph to flow in that direction

1

(*c*) Describe **one** way in which the composition of lymph differs from plasma.

it contains ~~little~~ less or no protein / no minerals / more fats

1

(*d*) What eventually happens to the lymph after it leaves the gland?

is entered into blood ✓

1

(*e*) Describe **one** function of the lymphatic system, apart from protecting the body from infection.

transports fats lipids back into blood. ✓

1

DO NOT WRITE IN THIS MARGIN

Marks

4. An investigation was carried out into the effect of caffeine on blood pressure, using coffee as the source of caffeine.

The systolic and diastolic blood pressures of six students were measured using a digital sphygmomanometer. Each student was then given a cup of coffee to drink. After one hour their blood pressure was measured again.

The results are shown in the table below.

Student	*Initial blood pressure* (mmHg)	*Final blood pressure* (mmHg)
1	120/75	134/82
2	127/79	145/88
3	118/70	124/72
4	134/81	143/83
5	122/73	133/77
6	129/84	137/90
Average	**125/77**	136/82

(*a*) Calculate the average final blood pressure and write your answer in the table above.

Space for calculation

1

(*b*) What conclusion can be drawn from these results?

That caffine increases blood pressure

1

(*c*) Describe an appropriate control for this investigation.

Do the same experiment with decaffinated coffee

1

(*d*) Apart from leaving one hour between readings, list **two** other variables which would need to be kept constant during this investigation.

1 volume of coffee drank

2 time left before measuring blood pressure

1

DO NOT WRITE IN THIS MARGIN

Marks

4. (continued)

(*e*) What is meant by systolic and diastolic blood pressure?

Systolic the pressure of blood when the heart chamber is systolic (contracted)

Diastolic Opposite

2

(*f*) The graphs below show initial and final blood pressures of one of the students.

Graph 1 Initial Blood Pressure

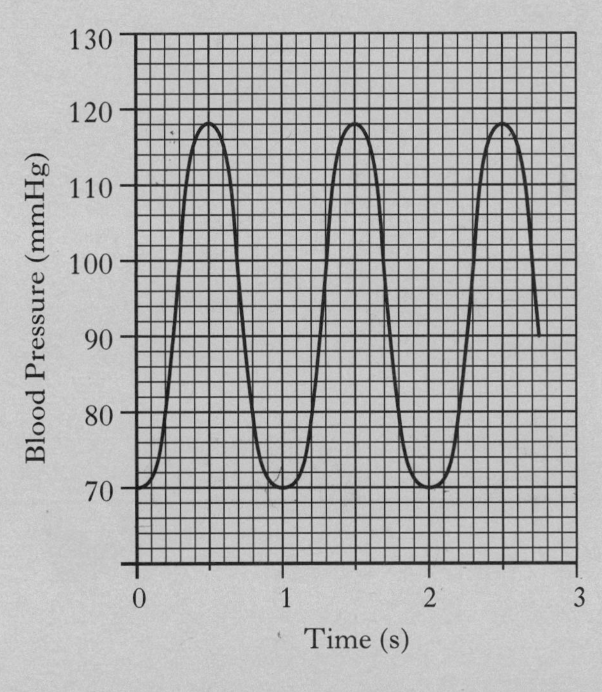

Graph 2 Final Blood Pressure

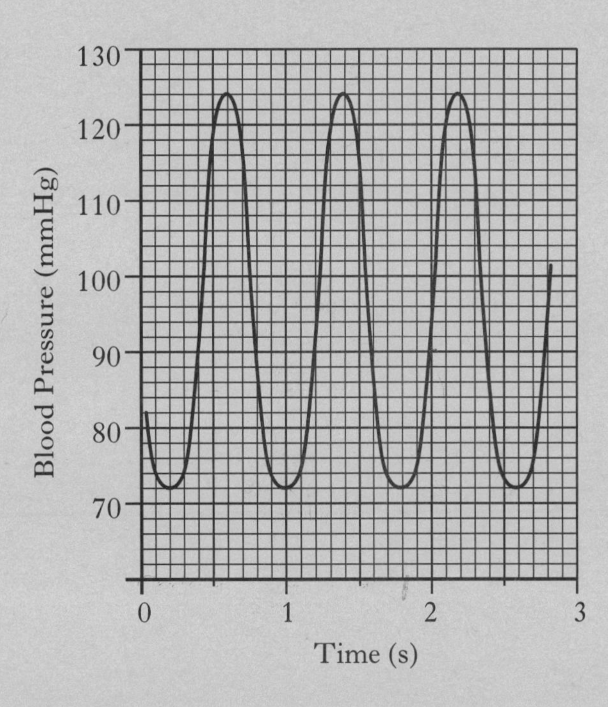

(i) Use the information in the table and the graphs to identify the student.

Student number 3

1

(ii) Calculate the increase in the pulse rate of this student over the period of the investigation.

Space for calculation

60

108 1·8

difference / initial value × 100

80 bpm

1

[Turn over

DO NOT WRITE IN THIS MARGIN

Marks

5. The diagram shows stages in the development of a human embryo from fertilisation to implantation.

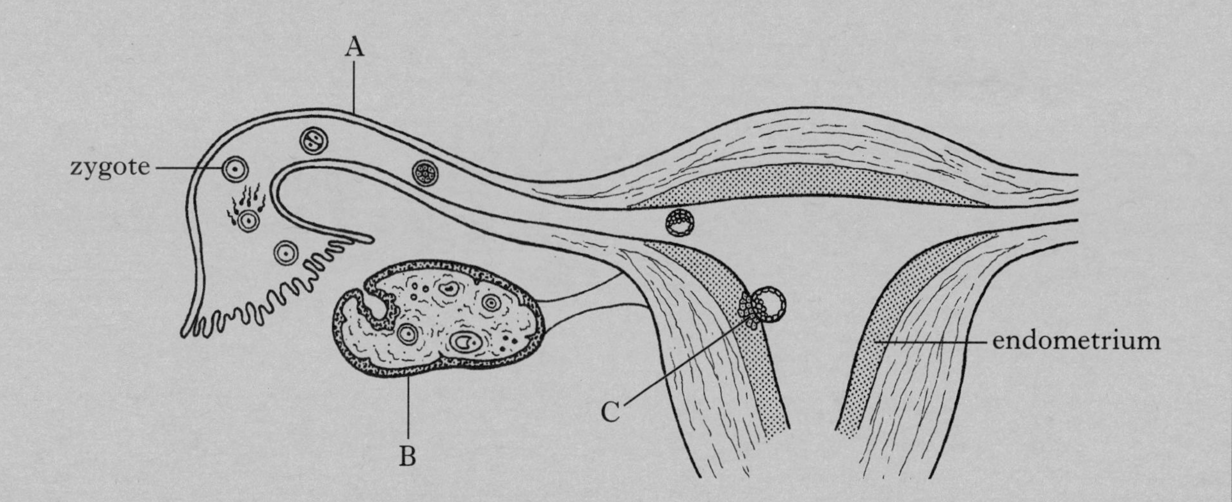

(*a*) Name the parts labelled **A** and **B**.

A oviduct B ovary **1**

(*b*) What term is used to describe the first few divisions of the zygote?

cleavage **1**

(*c*) Name a hormone which is involved in preparing the endometrium for implantation and state where it is produced.

Hormone progesterone Produced by corpus luteum **1**

(*d*) What organ will develop from the tissue labelled **C**?

placenta **1**

(*e*) Sometimes twins develop in the uterus. Distinguish between the formation of monozygotic and dizygotic twins.

monozygotic – one egg, one sperm
1 placenta, 2 embrionic sac one chorion

dizygotic – two eggs 2 sperm
two placenta, two embroinic sac one
cherion

3

DO NOT WRITE IN THIS MARGIN

Marks

6. The sperm counts of 30 men taken between 1940 and 2000 are shown in the graph below. A line of best-fit has been drawn, to indicate the trend over the 60-year period.

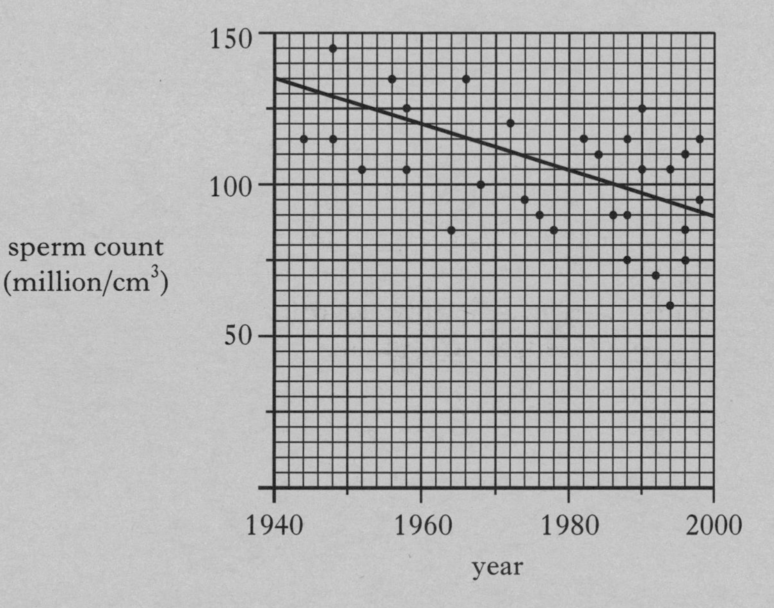

(*a*) Using the line of best-fit, calculate the percentage decline in sperm count over the 60-year period.

Space for calculation

____________ % **1**

(*b*) From the graph, what is the maximum sperm count for any one individual recorded during this period?

____________ million/cm^3 **1**

(*c*) Some insecticides are thought to influence sperm production. Explain why animals at the end of food chains are more likely to be affected by insecticides.

____

____

____ **1**

(*d*) Name the pituitary hormone which stimulates the production of sperm.

________________________ **1**

[Turn over

DO NOT WRITE IN THIS MARGIN

Marks

6. (continued)

(*e*) Name a gland which adds fluid to sperm during ejaculation and describe **one** function of this fluid.

Gland ____________________ **1**

Function of fluid ____________________

____________________ **1**

(*f*) Two treatments sometimes used for infertility are artificial insemination and *in vitro* fertilisation. Describe briefly what is meant by these terms.

artificial insemination ____________________

____

in vitro fertilisation ____________________

____________________ **2**

DO NOT WRITE IN THIS MARGIN

Marks

7. The Rhesus blood group system is determined by three pairs of alleles: **Cc**, **Dd** and **Ee**. However, only the **D** allele is important in blood transfusion and pregnancy. People with the dominant allele **D** are Rhesus positive and those with genotype **dd** are Rhesus negative.

(*a*) What term is used to describe characteristics controlled by many pairs of alleles?

______________________________ **1**

(*b*) Name another blood group system which has to be matched for blood transfusion to be successful.

______________________________ **1**

(*c*) What part of a cell carries the Rhesus antigen?

______________________________ **1**

(*d*) A Rhesus negative woman and a Rhesus positive man are planning to have a child. They consult a genetics counsellor to find out whether their child is likely to be Rhesus positive or Rhesus negative.

What genetic information could they be given?

____

____

____

____

____

______________________________ **2**

(*e*) Describe a treatment which can be used to protect a child at risk from the Rhesus reaction.

____

____

______________________________ **1**

[Turn over

DO NOT WRITE IN THIS MARGIN

Marks

8. The table below gives data on kidney transplants for the UK in the year 2001.

Category of patient	*Number*
On waiting list at beginning of 2001	**6052**
Received transplants during the year	
Removed from the list during the year	293
Died during the year	203
Added to the list during the year	1921
On waiting list at end of 2001	**6241**

(*a*) Complete the table by calculating the number of patients who received a transplant during the year 2001.

Space for calculation

1

(*b*) Assuming the same trend for the year 2002, predict the number of patients on the waiting list at the end of that year.

Space for calculation

____________ 1

(*c*) Drugs which suppress the immune system are given to transplant patients. Explain how this treatment reduces the chance of rejection of the transplanted kidney.

2

(*d*) Name the **two** blood vessels which would have to be cut and reconnected during a kidney operation.

____________ and ____________ 1

DO NOT WRITE IN THIS MARGIN

Marks

9. The diagram below is of a motor homunculus which represents the relative sizes of parts of the brain associated with motor control.

(*a*) In which part of the brain is the motor area located?

______________________________ **1**

(*b*) What is the function of the motor area of the brain?

______________________________ **1**

(*c*) Explain why the hands have such a large area of the brain devoted to their control in comparison to the feet.

____

______________________________ **1**

(*d*) What type of neural pathway is used to co-ordinate movements of the fingers?

______________________________ **1**

(*e*) Three facial expressions are shown below.

What term describes this type of communication?

______________________________ **1**

DO NOT WRITE IN THIS MARGIN

Marks

10. The diagram shows a neuromuscular junction.

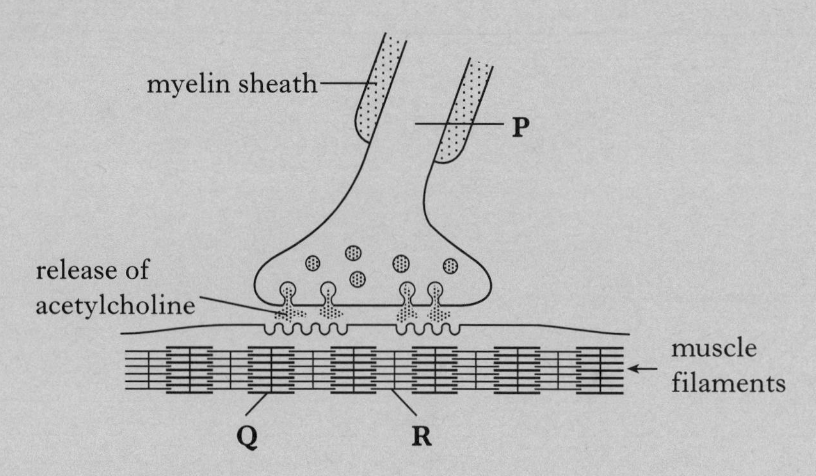

(*a*) Name the part of the nerve cell labelled **P**.

______________________________ **1**

(*b*) (i) What kind of substance is acetylcholine?

______________________________ **1**

(ii) What triggers the release of acetylcholine?

____

______________________________ **1**

(iii) State what happens to acetylcholine after it has acted on the muscle.

____

______________________________ **1**

(*c*) Name the **two** muscle proteins labelled **Q** and **R**.

Q ______________________________ **R** ______________________________ **1**

(*d*) Describe what happens to these protein filaments when a muscle contracts.

____

______________________________ **1**

DO NOT WRITE IN THIS MARGIN

Marks

11. Cod have been caught off the coast of Scotland for many years. The graph below shows the estimated population of cod in an area of the North Sea over the last hundred years.

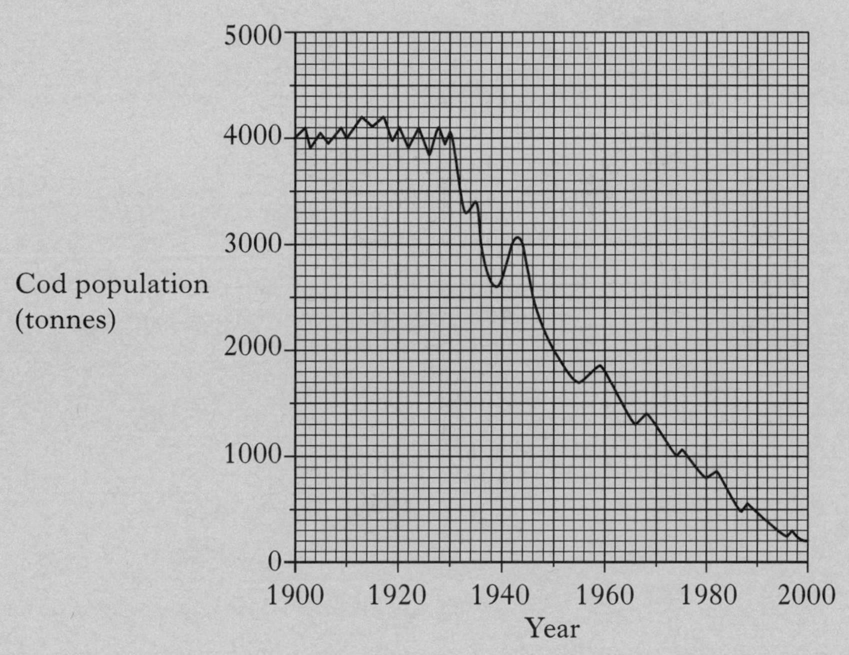

(*a*) Between the years 1900 and 1930 this area of the North Sea was at its carrying capacity for the cod population.

Explain what is meant by the term "carrying capacity".

____

____ **1**

(*b*) (i) Express, as a simple whole number ratio, the size of the cod population in 1950 to its size in 2000.

Space for calculation

______ : ______ **1**

(ii) Suggest **two** reasons for the decline of the cod population over the last 50 years.

1 ____

____

2 ____

____ **2**

[Turn over

DO NOT WRITE IN THIS MARGIN

Marks

12. The graphs show the changes in birth and death rates of two countries **A** and **B**.

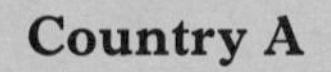

Country A

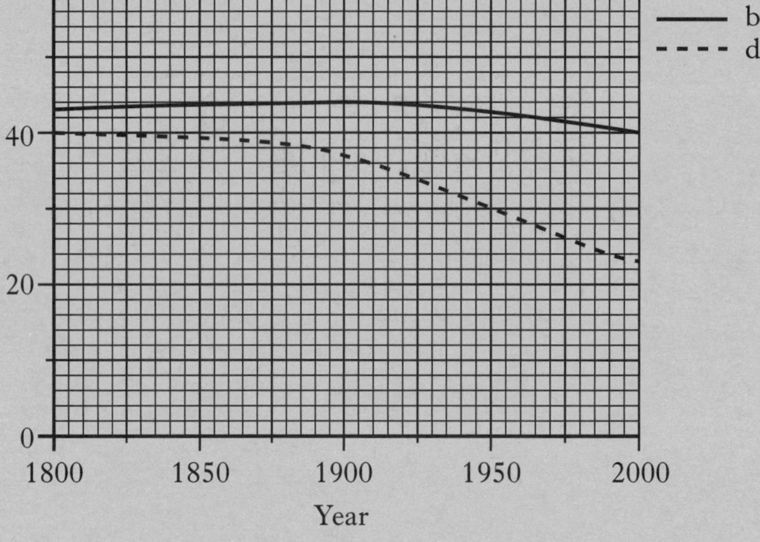

Country B

60
40
20
0
Number per 1000 of population
1800
1850
1900
1950
2000
Year

(*a*) (i) State the birth rate and death rate in country **A** in the year 1900.

Birth rate ______________ Death rate ______________ number per 1000 **1**

(ii) Suggest **one** reason for the decline in birth rate of country **A** over the last fifty years.

____

____ **1**

(iii) Suggest **one** reason for the decline in death rate of country **A** over the last fifty years.

____

____ **1**

DO NOT WRITE IN THIS MARGIN

12. (continued) *Marks*

(*b*) Is the population of country **B** increasing or decreasing over the period of time shown?

Give a reason for your answer.

Change ____________________

Reason ____

____

____ **1**

(*c*) Which of the **two** countries is likely to be a developing country?

Give a reason for your answer.

Country ____________________

Reason ____

____

____ **1**

[Turn over

DO NOT WRITE IN THIS MARGIN

Marks

13. An investigation was carried out into the effect of lack of nitrates on the growth of cress seedlings. Two equal batches of seeds were grown in agar gel containing:

A all necessary mineral salts

B all mineral salts except nitrate.

Each seed was placed on the surface of the agar in a glass tube as shown in the diagram below.

The heights of the seedlings were measured every day for eight days and the results are shown in the table.

Day	*Average height of seedlings in gel A* (mm)	*Average height of seedlings in gel B* (mm)
0	0	0
1	1	1
2	3	3
3	6	4
4	9	5
5	13	6
6	18	7
7	23	8
8	29	8

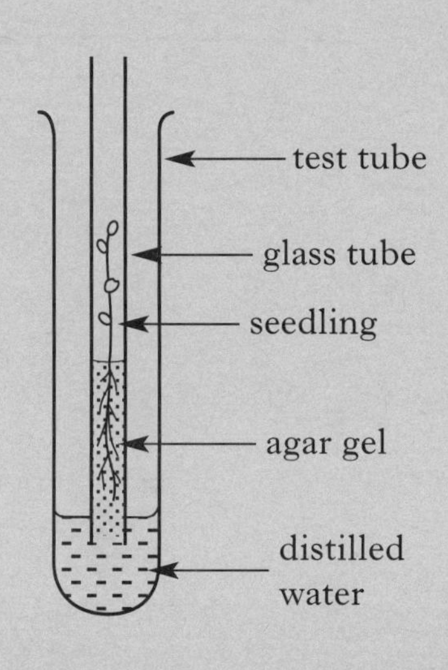

(*a*) Construct a line graph to illustrate the data in the table.

(Additional graph paper, if required, can be found on page 32.)

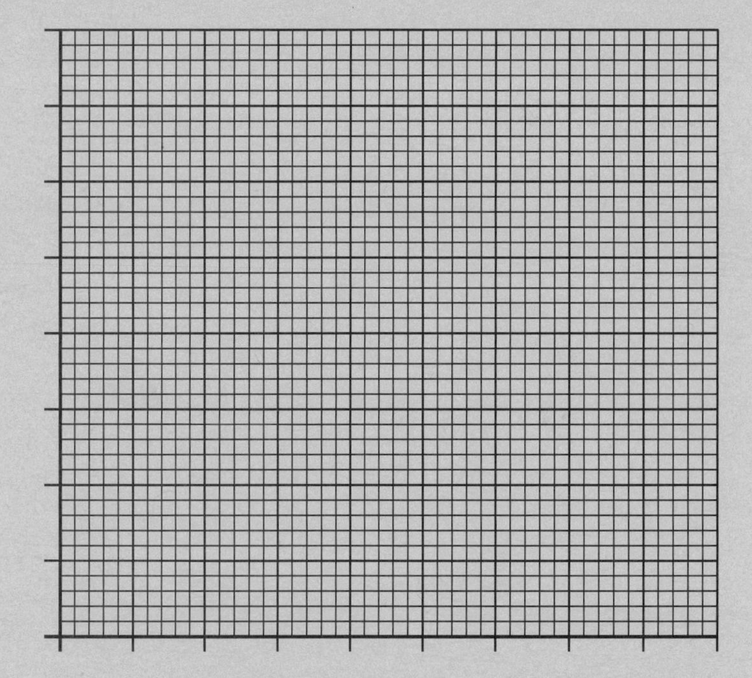

3

(*b*) State **one** conclusion which can be drawn from these results.

____

____ **1**

DO NOT WRITE IN THIS MARGIN

Marks

13. (continued)

(*c*) Identify the control and explain your choice.

Control ______________________

Explanation ______________________

______________________ **1**

(*d*) What feature of this investigation makes the results reliable?

______________________ **1**

(*e*) Suggest why distilled water is used in the test-tube rather than tap water.

____

______________________ **1**

(*f*) Phosphates are also necessary for good plant growth. Name **one** compound, other than ADP and ATP, which contains phosphate.

______________________ **1**

[Turn over

SECTION C

Marks

Both questions in this section should be attempted.

Note that each question contains a choice.

Questions 1 and 2 should be attempted on the blank pages which follow.

Supplementary sheets, if required, may be obtained from the invigilator.

Labelled diagrams may be used where appropriate.

1. Answer **either** A **or** B.

A. Give an account of memory under the following headings:

(i) short term memory; **4**

(ii) methods of transfer to long term memory. **6**

(10)

OR

B. Describe ways in which food production has been increased in the last fifty years under the following headings:

(i) land use; **4**

(ii) the use of chemicals. **6**

(10)

In question 2 ONE mark is available for coherence and ONE mark is available for relevance.

2. Answer **either** A **or** B.

A. Describe the events in meiosis which give rise to variation in gametes. **(10)**

OR

B. Describe how proteins are assembled from the code on a mRNA strand. **(10)**

[*END OF QUESTION PAPER*]

DO NOT WRITE IN THIS MARGIN

SPACE FOR ANSWERS

DO NOT WRITE IN THIS MARGIN

SPACE FOR ANSWERS

DO NOT WRITE IN THIS MARGIN

SPACE FOR ANSWERS

DO NOT WRITE IN THIS MARGIN

SPACE FOR ANSWERS

DO NOT WRITE IN THIS MARGIN

SPACE FOR ANSWERS

DO NOT WRITE IN THIS MARGIN

SPACE FOR ANSWERS

ADDITIONAL GRAPH PAPER FOR QUESTION 13(*a*)

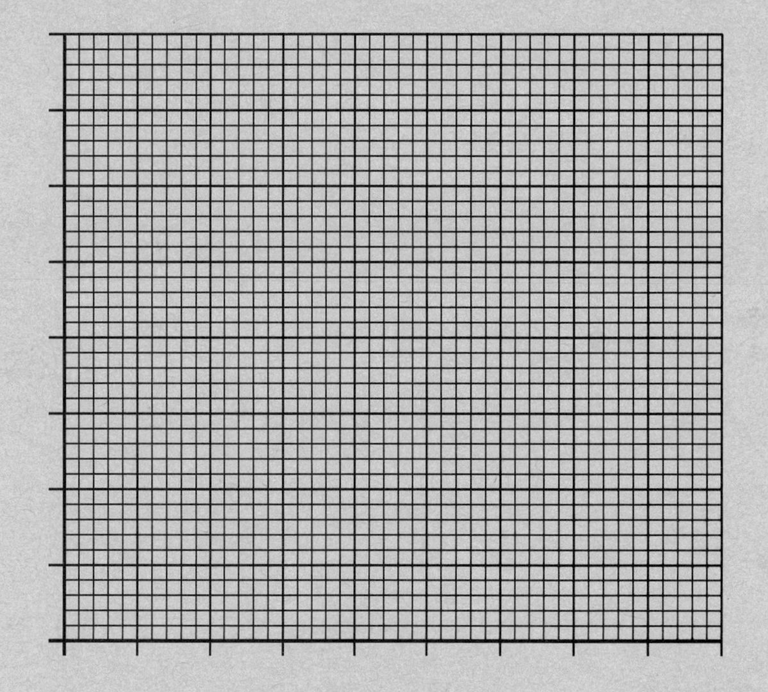

[BLANK PAGE]

FOR OFFICIAL USE

Total for Sections B & C

X009/301

NATIONAL QUALIFICATIONS 2005

WEDNESDAY, 18 MAY
1.00 PM – 3.30 PM

HUMAN BIOLOGY
HIGHER

Fill in these boxes and read what is printed below.

Full name of centre

Town

Forename(s)

Surname

Date of birth
Day Month Year

Scottish candidate number

Number of seat

SECTION A—Questions 1–30

Instructions for completion of Section A are given on page two.

SECTIONS B AND C

1 (a) All questions should be attempted.
  (b) It should be noted that in **Section C** questions 1 and 2 each contain a choice.

2 The questions may be answered in any order but all answers are to be written in the spaces provided in this answer book, and must be written clearly and legibly in ink.

3 Additional space for answers will be found at the end of the book. If further space is required, supplementary sheets may be obtained from the invigilator and should be inserted inside the **front** cover of this book.

4 The numbers of questions must be clearly inserted with any answers written in the additional space.

5 Rough work, if any should be necessary, should be written in this book and then scored through when the fair copy has been written. If further space is required a supplementary sheet for rough work may be obtained from the invigilator.

6 Before leaving the examination room you must give this book to the invigilator. If you do not, you may lose all the marks for this paper.

Read carefully

1 Check that the answer sheet provided is for **Human Biology Higher (Section A)**.

2 Check that the answer sheet you have been given has **your name**, **date of birth**, **SCN** (Scottish Candidate Number) and **Centre Name** printed on it.

Do not change any of these details.

3 If any of this information is wrong, tell the Invigilator immediately.

4 If this information is correct, **print** your name and seat number in the boxes provided.

5 Use **black** or **blue ink** for your answers. **Do not use red ink**.

6 The answer to each question is **either** A, B, C or D. Decide what your answer is, then put a horizontal line in the space provided (see sample question below).

7 There is **only one correct** answer to each question.

8 Any rough working should be done on the question paper or the rough working sheet, **not** on your answer sheet.

9 At the end of the exam, put the **answer sheet for Section A inside the front cover of this answer book**.

Sample Question

The digestive enzyme pepsin is most active in the

A mouth

B stomach

C duodenum

D pancreas.

The correct answer is **B**—stomach. The answer **B** has been clearly marked with a horizontal line (see below).

A B C D

Changing an answer

If you decide to change your answer, cancel your first answer by putting a cross through it (see below) and fill in the answer you want. The answer below has been changed to **B**.

A B C D

If you then decide to change back to an answer you have already scored out, put a tick (✓) to the **right** of the answer you want, as shown below:

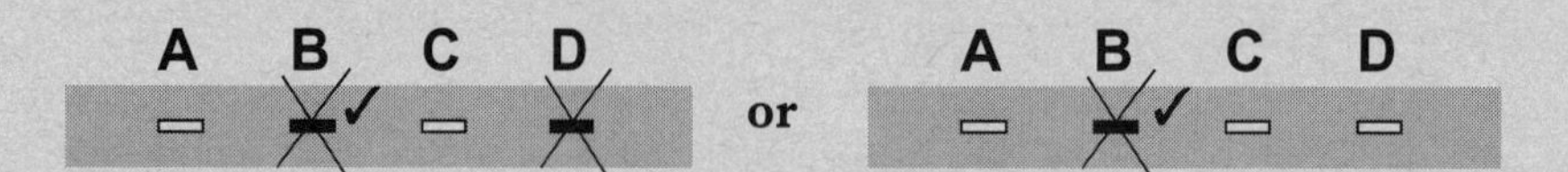

SECTION A

All questions in this section should be attempted.

Answers should be given on the separate answer sheet provided.

1. A series of enzyme-controlled reactions is shown below.

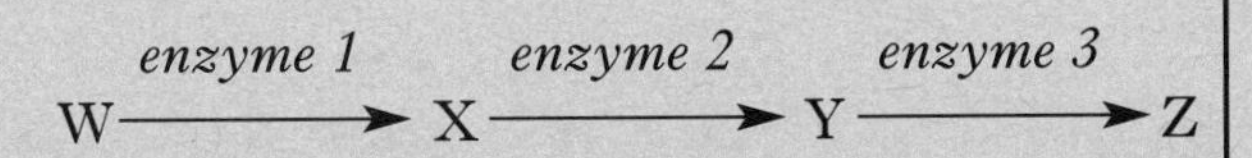

If an inhibitor which affects enzyme 2 is introduced to the system, which of the following will happen?

A X will accumulate

B Y will accumulate

C X and Y will accumulate

D Y and Z will accumulate

2. Which of the following describes metabolism correctly?

A The breakdown of chemicals to release energy

B The synthesis of large molecules

C The chemical reactions of organisms

D The breakdown of food molecules

3. The following diagram shows part of a protein molecule.

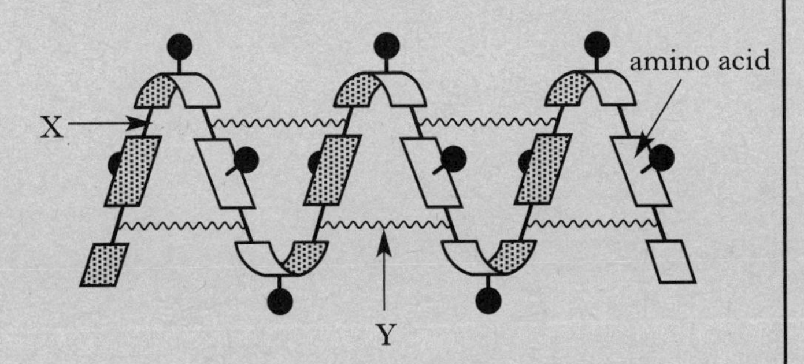

Which line in the table describes correctly bonds X and Y?

	Bond X	*Bond Y*
A	hydrogen	peptide
B	hydrogen	hydrogen
C	peptide	hydrogen
D	peptide	peptide

Questions 4 and 5 refer to muscle filaments.

4. Which line of the table identifies correctly the types of filaments found in the light and dark bands of striated muscle?

	Banding pattern	
	Light	*Dark*
A	actin	myosin
B	myosin	actin + myosin
C	myosin	actin
D	actin	actin + myosin

5. When a muscle contracts what happens to these filaments?

A Both filaments contract

B Actin contracts but not myosin

C Myosin contracts but not actin

D The filaments slide over one another

6. The diagram of the cell is magnified 400 times. What is the true size of the cell?

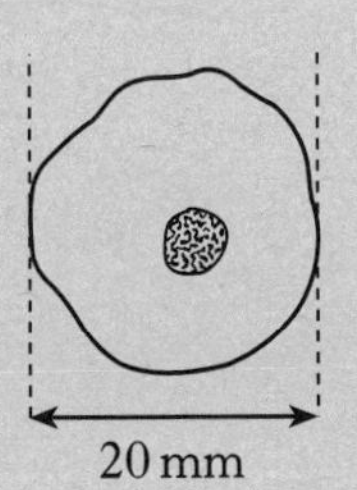

A 20 μm

B 50 μm

C 80 μm

D 500 μm

[Turn over

7. Which of the following statements is true of all viruses?

A They have a protein-lipid coat and contain DNA.

B They have a protein-lipid coat and contain RNA.

C They have a protein coat and a nucleus.

D They have a protein coat and contain nucleic acid.

8. How many adenine molecules are present in a DNA molecule of 2000 bases, if 20% of the base molecules are cytosine?

A 200

B 300

C 400

D 600

9. In the formation of gametes, when does DNA replication occur?

A Before the start of meiosis

B As homologous chromosomes pair up

C At the end of the first meiotic division

D At the separation of chromatids

10. Alleles can be described as

A opposite types of gamete

B different versions of a gene

C identical chromatids

D non-homologous chromosomes.

11. A person has blood group AB.

Which entry on the table identifies correctly the antigens and antibodies present?

	Antigens on cells	*Antibodies in plasma*
A	A and B	anti-A and anti-B
B	none	anti-A and anti-B
C	A and B	none
D	none	none

12. The gene for albinism is autosomal and recessive. A couple who are both carriers of the gene have a son. What is the chance that he will have the same genotype as his parents?

A 1 in 1

B 1 in 2

C 1 in 3

D 1 in 4

13. The family tree below shows the transmission of the Rhesus D-antigen. The gene for the Rhesus D-antigen is not sex-linked.

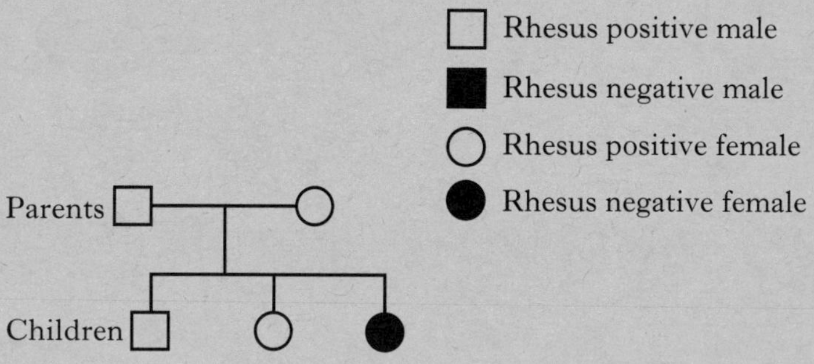

The parents are expecting a fourth child.

What is the chance that this child will be Rhesus negative?

A 0%

B 25%

C 50%

D 100%

14. Colour blindness is a sex-linked recessive trait.

A woman would have a 50% chance of being colour blind if

A both of her parents are carriers

B her father has normal vision but her mother is a carrier

C her father is a carrier and her mother is colour blind

D her father is colour blind and her mother is a carrier.

15. Which of the following may result in the presence of an extra chromosome in the cells of a human being?

A Non-disjunction

B Crossing over

C Segregation

D Inversion

16. As an ovum develops within the ovary, it is surrounded by

A a Graafian follicle

B seminal fluid

C a corpus luteum

D the endometrium.

17. Which line in the table best describes dizygotic twins?

	Number of sperm involved in formation	*Number of ova involved in formation*	*Resulting genotypes*
A	1	1	identical
B	1	1	non-identical
C	2	2	identical
D	2	2	non-identical

18. The diagram below illustrates the hormonal control of a 30-day menstrual cycle.

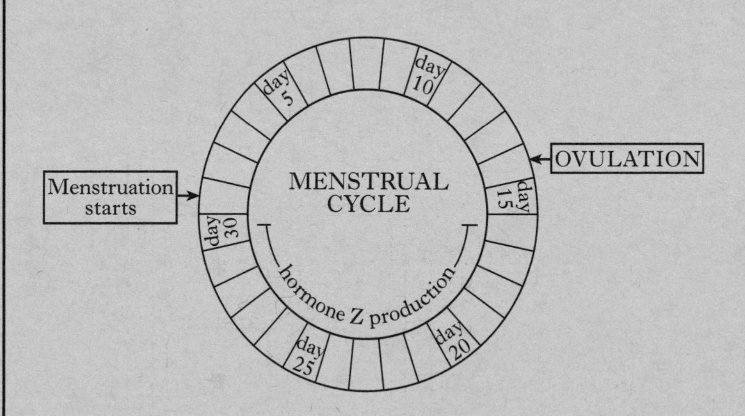

Which line of the table identifies correctly hormone Z and the structure which produces this hormone?

	Hormone Z	*produced by*
A	LH	ovary
B	oestrogen	corpus luteum
C	progesterone	Graafian follicle
D	progesterone	corpus luteum

19. Which of the following babies would be most likely to require a blood transfusion immediately after birth?

A The first baby of a Rhesus negative mother and Rhesus positive father

B The first baby of a Rhesus positive mother and Rhesus negative father

C The second baby of a Rhesus negative mother and Rhesus positive father

D The second baby of a Rhesus positive mother and Rhesus negative father

[Turn over

20. The diagram below shows the relationship between blood capillaries, body cells and lymph capillaries.

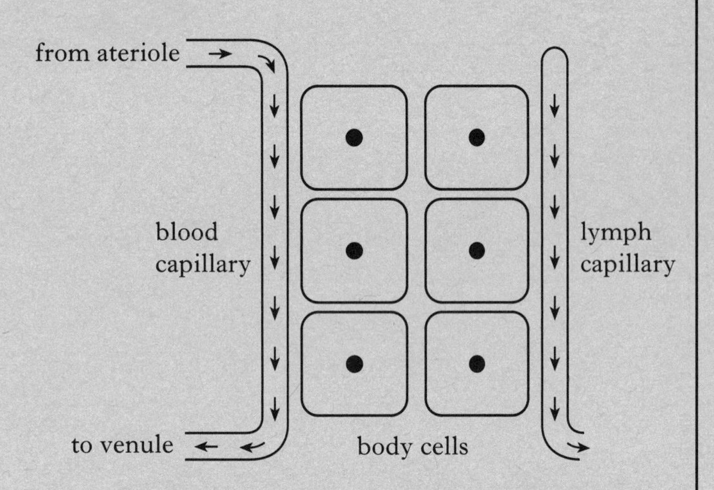

Which of the following is a correct description of the movement of oxygen to and from the body cells?

A From body cells to blood and lymph capillaries

B From blood capillaries to body cells

C From lymph capillaries to body cells

D From blood and lymph capillaries to body cells

21. If body temperature drops below normal, which of the following would result?

A Vasodilation of skin capillaries

B Vasoconstriction of skin capillaries

C Decreased metabolic rate

D Increased sweating

22. The graphs below show the average yearly increase in height of girls and boys.

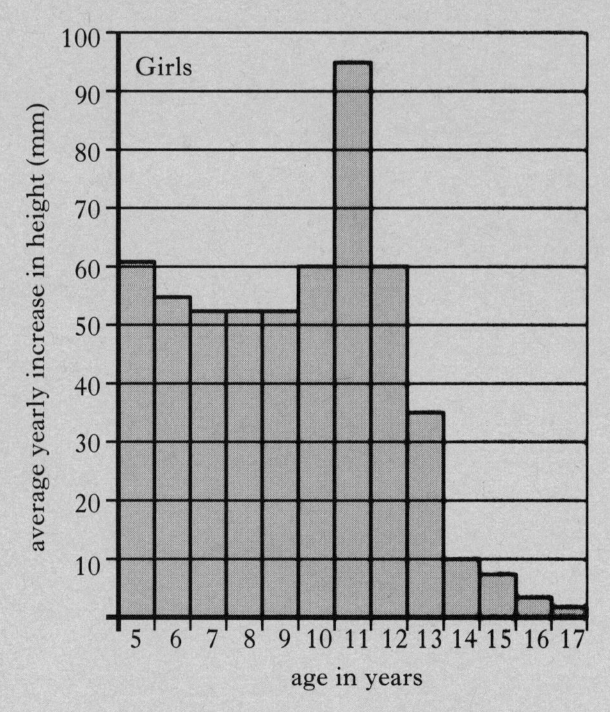

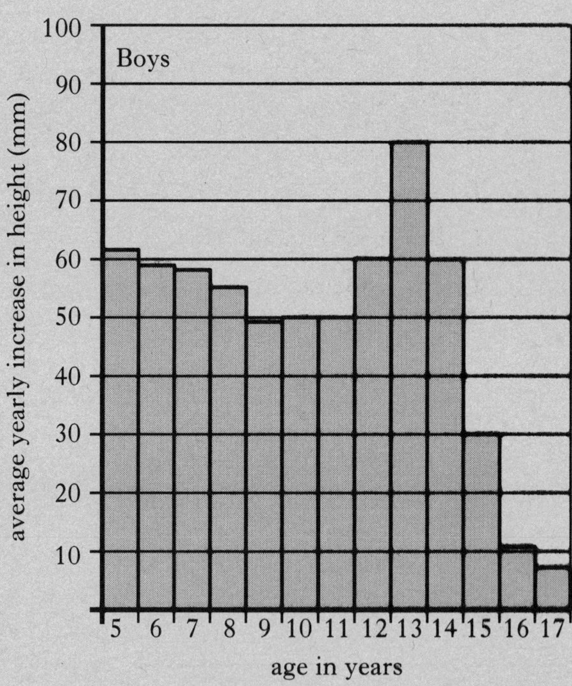

Which of the following statements is correct?

A The greatest average yearly increase for boys occurs one year later than the greatest average yearly increase for girls.

B Boys are still growing at seventeen but girls have stopped growing by this age.

C Between the ages of five and eight boys grow more than girls.

D There is no age when boys and girls show the same average yearly increase in height.

23. The diagram below shows a motor neurone.

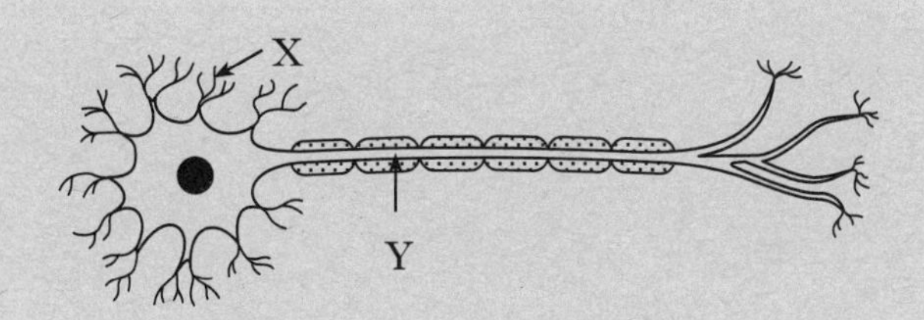

Which line of the table identifies correctly the labelled parts and the direction of impulse?

	X	*Y*	*Direction*
A	dendrite	axon	X → Y
B	dendrite	axon	Y → X
C	axon	dendrite	X → Y
D	axon	dendrite	Y → X

24. Vision in dim light is improved by the rods having

A peripheral neural pathways

B diverging neural pathways

C central neural pathways

D converging neural pathways.

25. The histogram shows the percentage distribution of IQ rating in a sample of 1000 Scottish children.

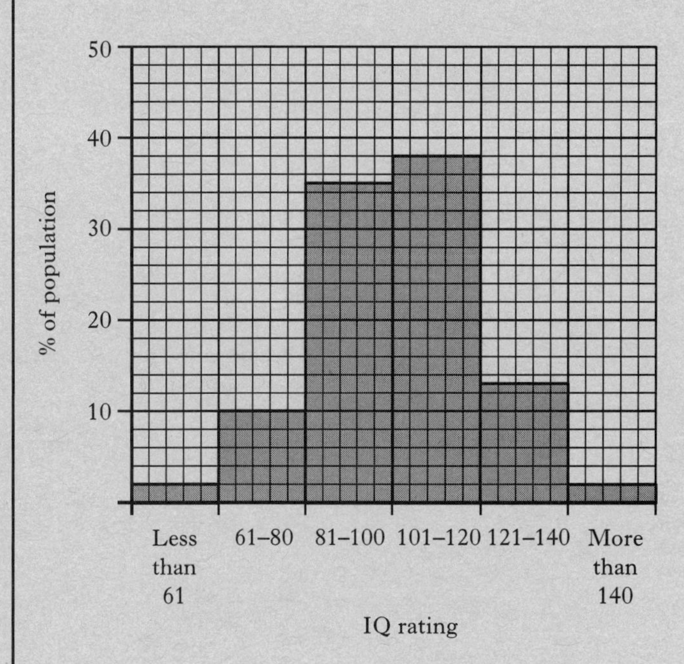

How many children have an IQ of over 120?

A 15

B 53

C 150

D 530

26. Students were asked to recall twelve letters of the alphabet in any order, after hearing the list of letters read slowly once over. An analysis of their performance is shown in the graph below.

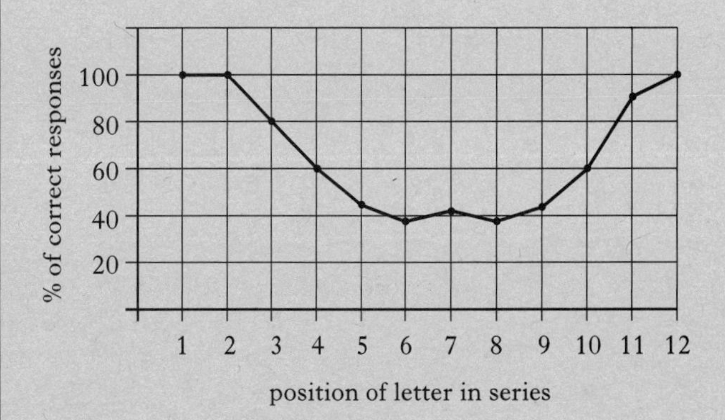

On how many occasions was a letter recalled by more than half of the students?

A 5

B 7

C 9

D 10

27. Rivers polluted by raw sewage have low oxygen concentrations as a direct result of

A large numbers of bacteria

B algal blooms

C high nutrient levels

D low nutrient levels.

28. The diagram represents part of the nitrogen cycle.

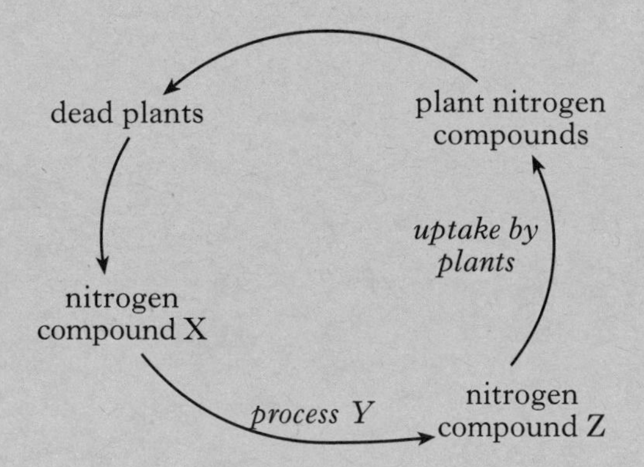

Process Y is the production of

A ammonia by decomposition

B nitrates by nitrification

C ammonia by nitrogen fixation

D nitrates by denitrification.

29. A country has a population of 10 million. What is the likely increase in population over a two-year period given a growth rate of 2% per annum?

A 102 000

B 104 040

C 204 000

D 404 000

30. The diagram below shows a population pyramid for a country.

Males | Age | Females

75+
70 – 74
65 – 69
60 – 64
55 – 59
50 – 54
45 – 49
40 – 44
35 – 39
30 – 34
25 – 29
20 – 24
15 – 19
10 – 14
5 – 9
0 – 4

8 6 4 2 0 | 0 2 4 6 8

population size (millions)

How many girls between the ages of 10 and 19 are there in the population?

A 6 million

B 10 million

C 12 million

D 21 million

Candidates are reminded that the answer sheet MUST be returned INSIDE the front cover of this answer booklet.

[Turn over for Section B]

DO NOT WRITE IN THIS MARGIN

Marks

SECTION B

All questions in this section should be attempted.

1. The diagram below represents the process of RNA synthesis.

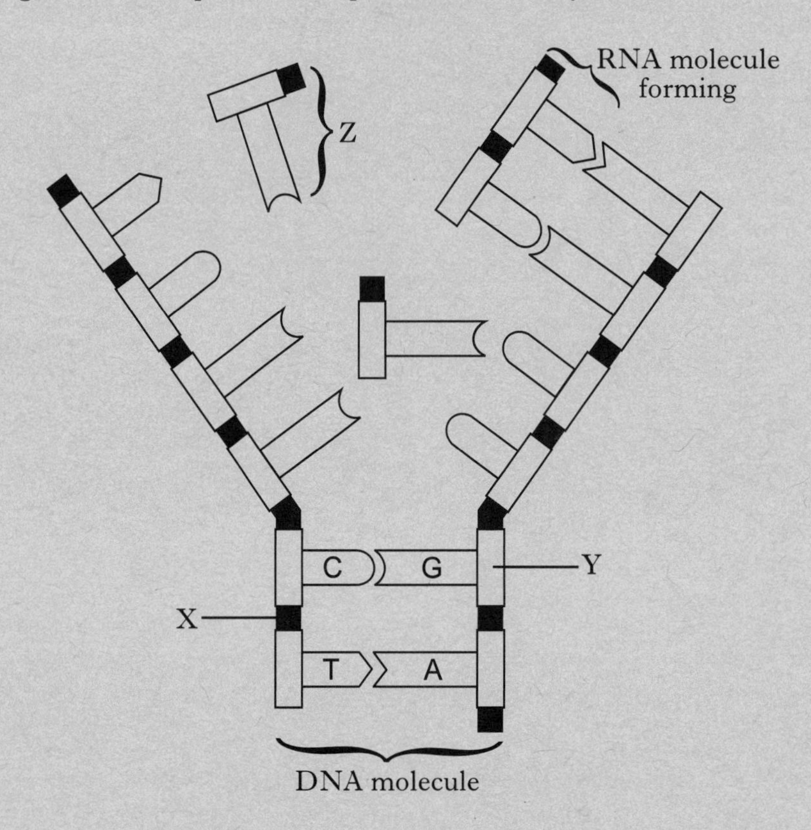

(*a*) Where in the cell does this process take place?

________________________________ **1**

(*b*) Name the components X, Y and Z.

X ______________________________

Y ______________________________

Z ______________________________ **2**

(*c*) State the full names of any **two** different RNA bases shown in the diagram.

1 ______________________________

2 ______________________________ **1**

(*d*) Name another substance, **not** shown in the diagram, which is essential for RNA synthesis.

________________________________ **1**

DO NOT WRITE IN THIS MARGIN

1. (continued) *Marks*

(*e*) (i) What name is given to the triplets of bases in an mRNA molecule?

_______________________________________ **1**

(ii) The table below shows some amino acids and the triplets of bases specific to them.

Amino acid	*Triplet of mRNA bases*
alanine	GCU
arginine	CGA
serine	UCG
histidine	CAC
valine	GUG

Name the **two** amino acids that would be specified by the mRNA molecule forming on the DNA strand in the diagram.

1 ____________________________ 2 ____________________________ **1**

[Turn over

DO NOT WRITE IN THIS MARGIN

Marks

2. The diagram below shows three stages in the humoral immune response.

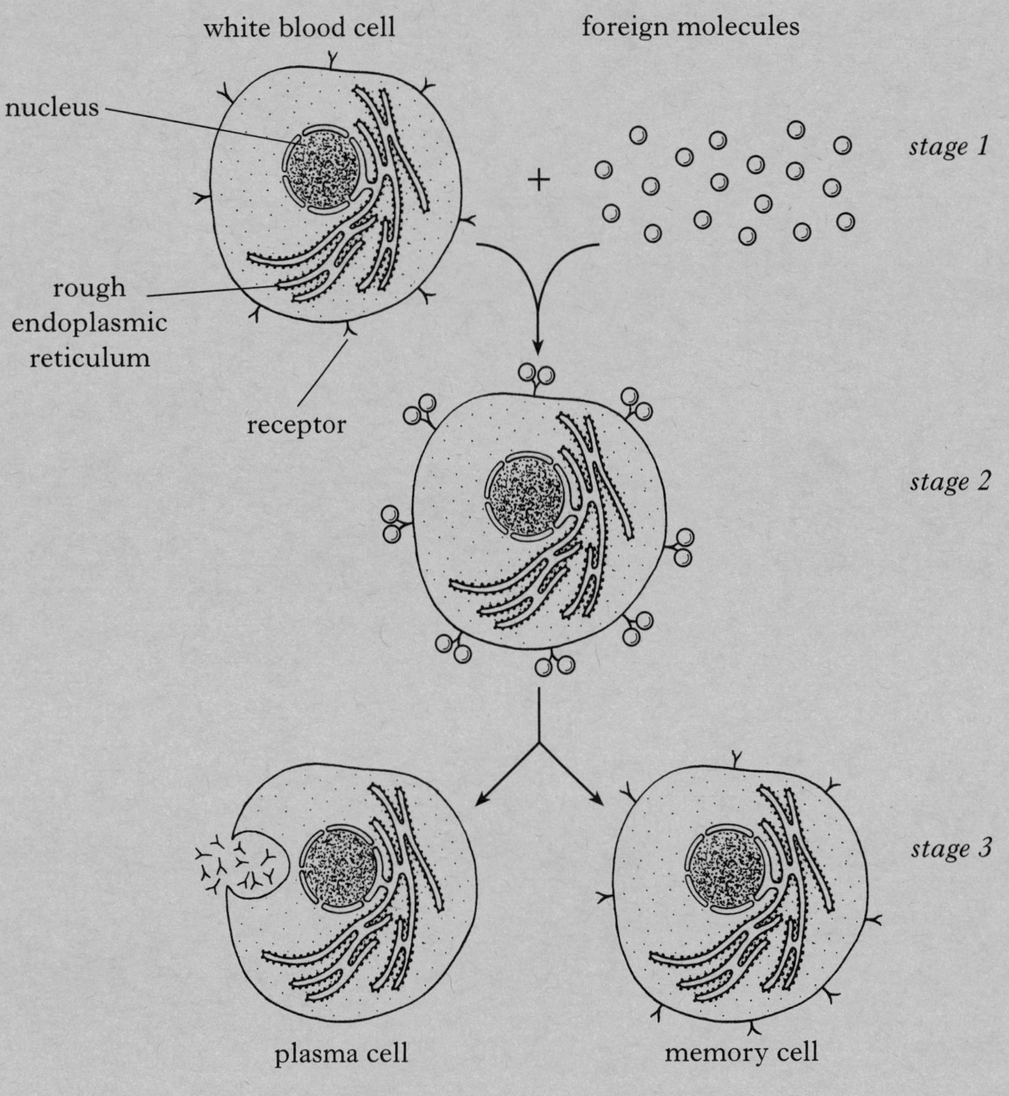

(*a*) (i) What type of white blood cell carries out the humoral immune response?

________________________ **1**

(ii) What name is given to foreign molecules which stimulate the immune response?

________________________ **1**

(*b*) Describe **two** responses made by the white blood cell as a result of the attachment of the foreign molecules.

1 ________________________

2 ________________________ **1**

DO NOT WRITE IN THIS MARGIN

2. (continued) *Marks*

(*c*) Mature plasma cells contain a large quantity of rough endoplasmic reticulum. Explain this feature of these cells.

____

____

______________________________ **2**

(*d*) Suggest the role of memory cells in the immune response.

____

______________________________ **1**

(*e*) What term describes the secretion of substances, such as antibodies, out of a cell?

______________________________ **1**

(*f*) Describe how the body might obtain antibodies in a natural, passive way.

____

______________________________ **1**

[Turn over

DO NOT WRITE IN THIS MARGIN

Marks

3. The diagram below summarises a metabolic pathway within a cell.

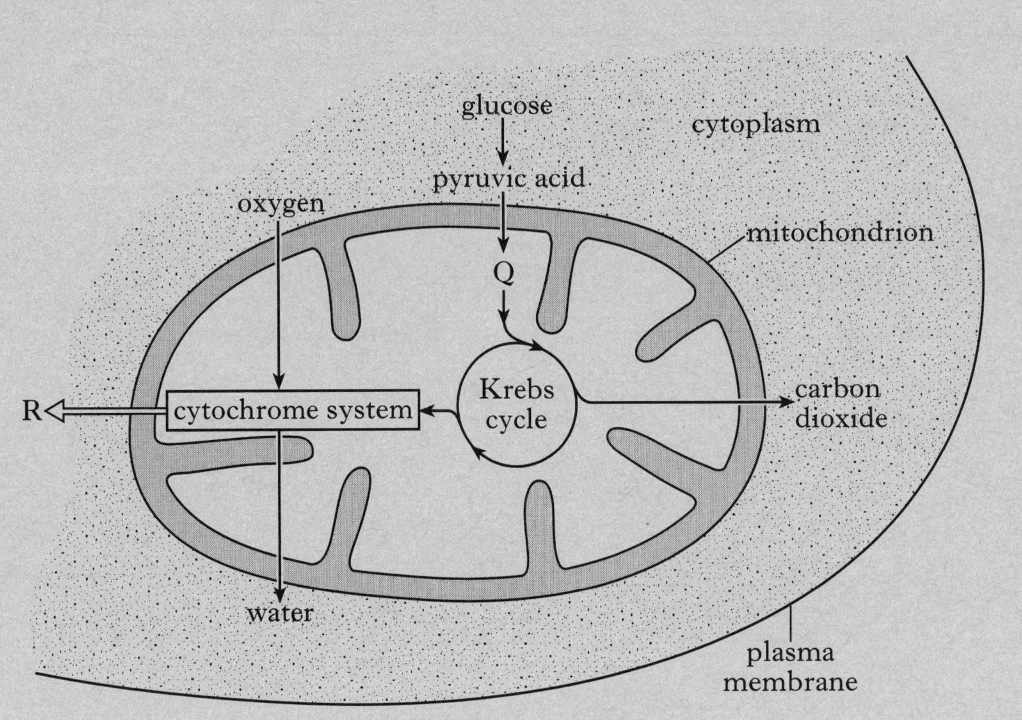

(*a*) Name the process which results in the formation of pyruvic acid.

______________________________ **1**

(*b*) Name substance Q.

______________________________ **1**

(*c*) How many carbon atoms are removed in one turn of the Krebs cycle?

______________ **1**

(*d*) What is the role of NAD in this process?

______________________________ **1**

(*e*) Why does the cytochrome system stop when oxygen is absent?

______________________________ **1**

(*f*) Substance R is the main product of the cytochrome system.

Where in this metabolic pathway is substance R required?

______________________________ **1**

DO NOT WRITE IN THIS MARGIN

Marks

4. The diagram shows sections of a testis and two seminiferous tubules.

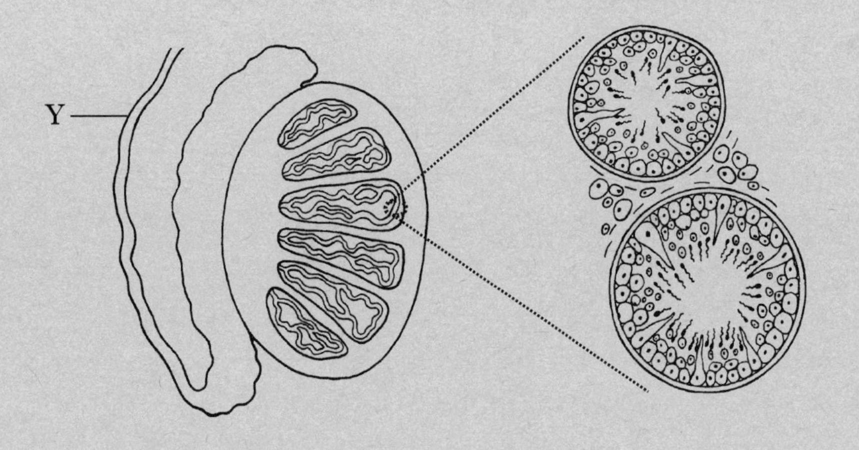

(*a*) Name structure Y. **1**

(*b*) Follicle stimulating hormone (FSH) affects the testes.

(i) State where FSH is produced in the body. **1**

(ii) What effect does FSH have on the testes? **1**

(*c*) (i) On **Section B** use an **X** to mark the site of testosterone production. **1**

(ii) Describe how the concentration of testosterone in the blood is prevented from becoming too high. **2**

(iii) Suggest why testosterone injections are sometimes used to treat infertility in men. **1**

DO NOT WRITE IN THIS MARGIN

Marks

5. A nomogram is shown below. Nomograms are used to estimate the surface area of individuals.

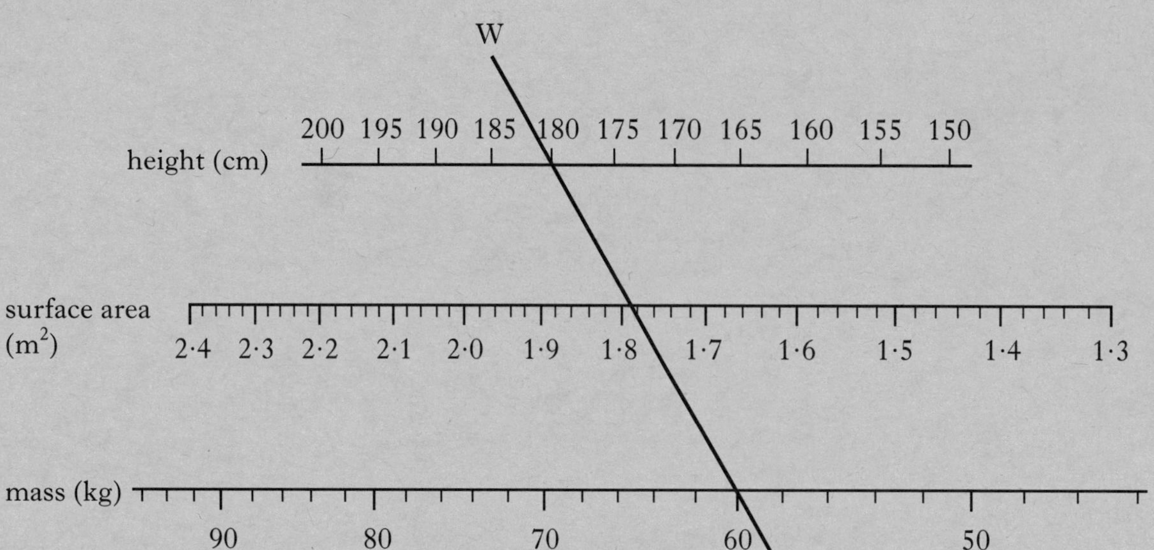

(*a*) The table below contains information about three individuals. Use the nomogram to complete the table. Line W has been completed as an example.

Individual	*Mass* (kg)	*Height* (cm)	*Surface Area* (m^2)
W	60	180	1·79
X	70	160	
Y	56		1·58

1

(*b*) The table below shows the surface area and volume of two boys.

Name	*Surface Area* (m^2)	*Volume* (dm^3)
Iain	2	50
Andy	2	60

Which of these boys is likely to be more susceptible to hypothermia?

Give a reason for your answer.

Boy ________________

Reason ____

____ **1**

(*c*) Name the microscopic structures (1) in the lungs and (2) in the small intestine, which provide an increased surface area.

1 Lungs ____________________________

2 Small intestine __________________________ **1**

DO NOT WRITE IN THIS MARGIN

Marks

6. The graph below shows the drop in pressure as blood flows through the circulatory system.

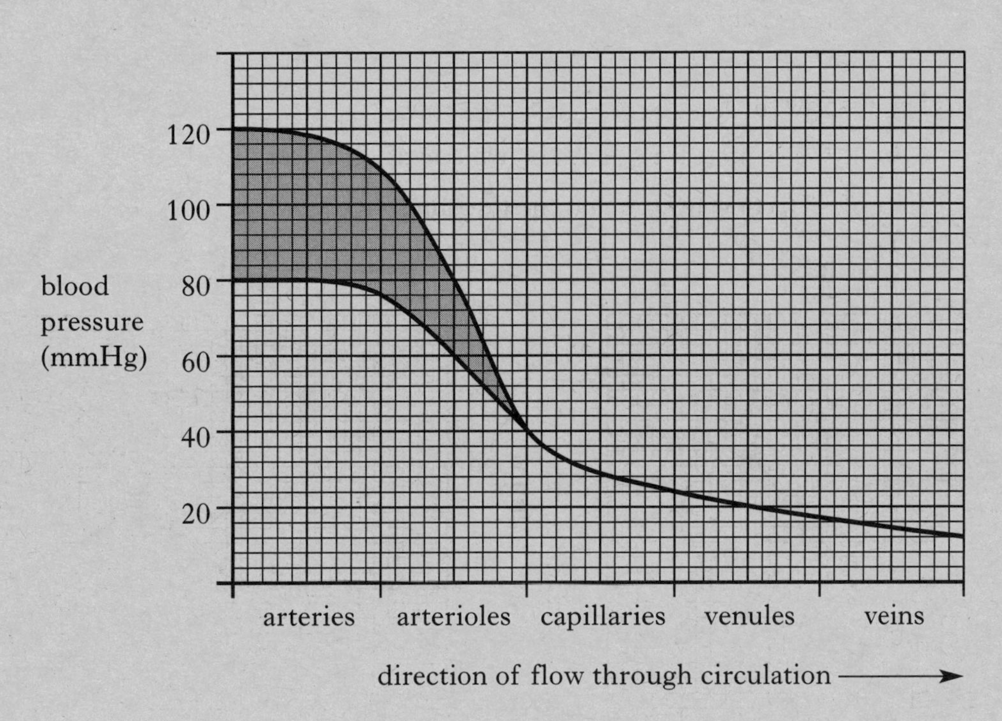

(*a*) Calculate the decrease in pressure that occurs in the capillaries.

Space for calculation

_____________ mmHg **1**

(*b*) The pressure of the blood is highest as it leaves the heart. Where in the circulation would blood be found at a pressure 25% of this value?

_____________________________ **1**

(*c*) Why is there a maximum and minimum value given for the arteries and arterioles?

____

____ **1**

(*d*) Name the blood vessels which link the following organs.

1 From brain to heart _____________________________

2 From small intestine to liver _____________________________

3 From heart to lungs _____________________________ **2**

[Turn over

DO NOT WRITE IN THIS MARGIN

Marks

7. An investigation was carried out to find out how the percentage concentration of carbon dioxide (CO_2) in inhaled air affects the volume of air breathed and the breathing rate. Ten subjects were chosen and tested at seven different concentrations of CO_2.

The graphs below show the results of this investigation.

Graph 1 Effect of CO_2 concentration on the volume of air inhaled

Graph 2 Effect of CO_2 concentration on the breathing rate

Graph 1

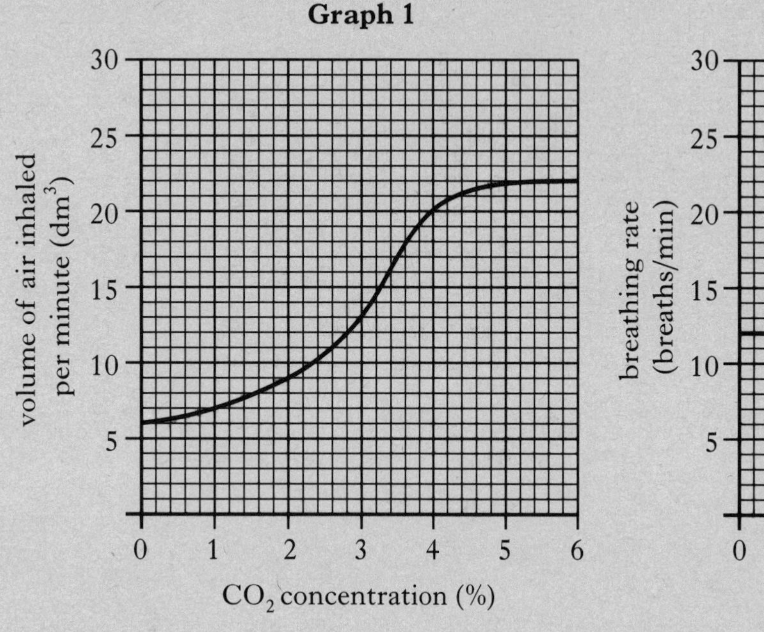

Graph 2

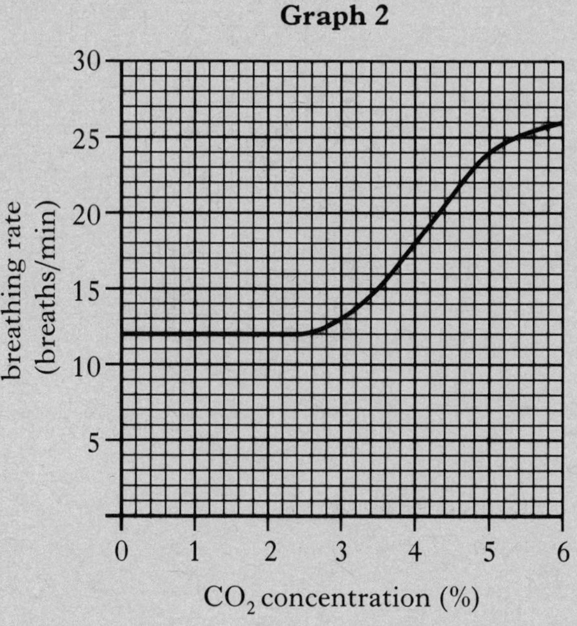

(*a*) From **Graph 1**, what is the volume of air inhaled in one minute when the CO_2 concentration is 3%?

________________ **1**

(*b*) From **Graph 2**, describe the effect of increasing CO_2 concentration on breathing rate.

____

____

________________________________ **2**

(*c*) (i) Complete the table below to show the mean volume of air inhaled in a single breath at each of the concentrations of CO_2 given.

CO_2 concentration of inhaled air (%)	*Volume of air inhaled per minute* (dm^3)	*Breathing rate* (breaths per minute)	*Mean volume of one breath* (dm^3)
0	6	12	0·50
2	9	12	
4	20		
6			

2

DO NOT WRITE IN THIS MARGIN

Marks

7. (c) (continued)

(ii) Draw a graph to show the relationship between the concentration of CO_2 in inhaled air and the mean volume of one breath.

(Additional graph paper, if required, can be found on page 32.)

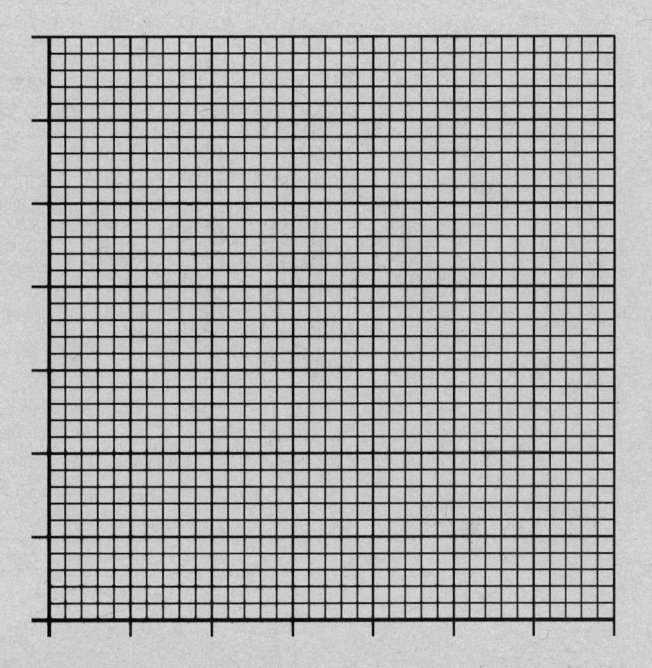

2

(iii) What conclusion can be drawn from the graph? Quote data from your graph to illustrate your answer.

____

______________________________ **1**

(d) (i) Before each reading was taken, each subject breathed the air samples for two minutes. Suggest a reason for this.

______________________________ **1**

(ii) Suggest another variable, apart from time, which would have to be controlled between each reading.

______________________________ **1**

(e) Suggest why ten subjects were chosen rather than just one.

____

______________________________ **1**

[Turn over

DO NOT WRITE IN THIS MARGIN

Marks

8. An investigation was carried out to determine the rates of flow and the composition of fluids in a human kidney. These were measured at positions P, Q, R and S, shown in the diagram below.

Kidney nephron

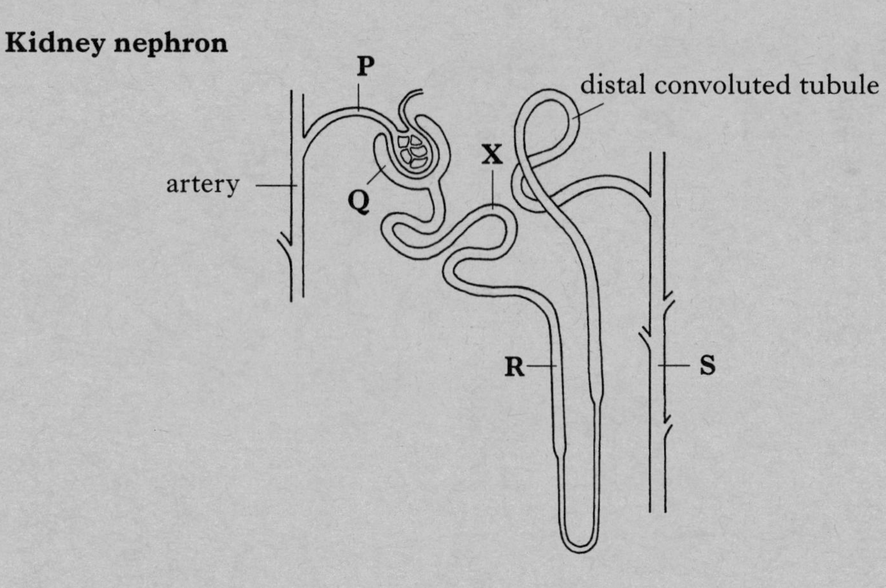

The results are given in the table below.

		Solute concentration (g/100 cm^3)		
Position	*Total flow rate through kidney* (cm^3/minute)	*Protein*	*Glucose*	*Urea*
P	1000	7·4	0·1	0·03
Q	100	0·0	0·1	0·03
R	20	0·0	0·0	0·15
S	1	0·0	0·0	1·85

(*a*) (i) Name structure **X**.

____________________ **1**

(ii) What process takes place in this part of the nephron?

____________________ **1**

(*b*) Explain why there is no protein at point **Q** in the nephron.

____

____________________ **1**

(*c*) (i) By how many times does the concentration of urea increase between points **Q** and **R**?

__________ **1**

DO NOT WRITE IN THIS MARGIN

Marks

8. (*c*) (continued)

(ii) Explain why the concentration of urea increases between points **R** and **S**.

____

____ **1**

(iii) Using data from the table, calculate the weight of urea which would pass from the collecting duct (**S**) to the bladder in one hour.

Space for calculation

____________ g **1**

(*d*) Express the concentration of glucose at point **Q** in grams per litre.

Space for calculation

____________ g/l **1**

(*e*) What effect would an increasing concentration of ADH in the blood have on each of the following?

(i) The concentration of urea at point **S**.

__________________________ **1**

(ii) The concentration of glucose at point **P**.

__________________________ **1**

[Turn over

DO NOT WRITE IN THIS MARGIN

Marks

9. The diagram shows the main parts of the human brain as seen in a vertical section.

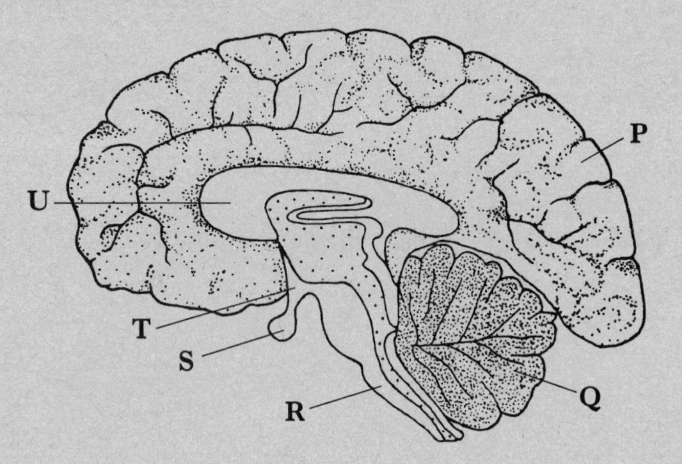

(*a*) Complete the table by adding the correct letters, names and functions of the parts.

Label	*Name*	*Function*
P		
	pituitary gland	
		temperature regulation

3

(*b*) Describe a feature of part **P** which improves its function.

____ 1

(*c*) What is meant by the term "localisation of function"?

____

____ 1

(*d*) Why is the part of the brain which controls the right hand much larger than the part which controls the right foot?

____

____

____ 1

DO NOT WRITE IN THIS MARGIN

Marks

10. The following diagrams represent a form of communication.

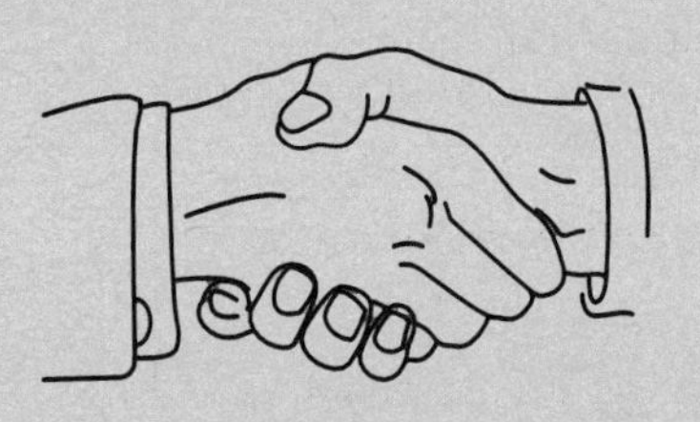
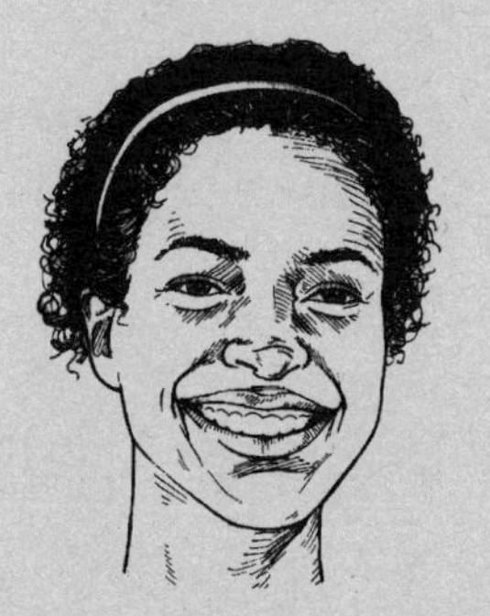
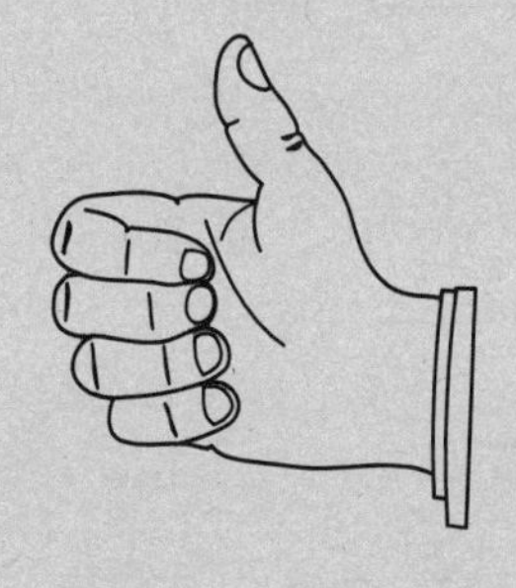

(*a*) What is this form of communication called?

______________________________ **1**

(*b*) (i) Describe the significance of such communication in infancy.

____

______________________________ **1**

(ii) Give **two** examples of this type of communication which are used by babies.

1 ______________________________

2 ______________________________ **1**

(*c*) The following are standard symbols.

Why are such symbols used worldwide?

____

______________________________ **1**

[Turn over

DO NOT WRITE IN THIS MARGIN

Marks

11. The account below relates to the effect of experience on behaviour.

> Nicky decided she wanted to learn how to play golf. Sam, the professional, was very helpful, offering her five introductory lessons at a reasonable rate, with the offer of five more if Nicky showed consistent improvement. He emphasised that she would have to pay careful attention to his demonstrations and copy his technique.
>
> Nicky enjoyed driving the ball but hated putting, so Sam always started lessons with putting and only moved on to using other clubs when sufficient improvement was shown. As the lessons went on, Sam expected a higher and higher standard before any driving was allowed.
>
> Four years later Nicky was good enough to represent her country at junior level but she refused to use any of the latest graphite-shafted clubs as she had lost her first championship when using a graphite putter.

(*a*) With reference to the above account, give an example of each of the following types of behaviour.

Imitation ______________________________

______________________________ **1**

Generalisation ______________________________

______________________________ **1**

(*b*) The professional used the technique of shaping in his teaching.

What is meant by "shaping"? Give an example from the text above.

Shaping ______________________________

______________________________ **1**

Example ______________________________

______________________________ **1**

(*c*) As well as rewarding Nicky for doing well, Sam could also have punished any poor performance. What term is used to describe this type of training?

______________________________ **1**

DO NOT WRITE IN THIS MARGIN

Marks

11. (continued)

(*d*) The paragraph below provides further information on human behaviour.

> Nicky loved playing in big championships as she found she always played better in front of a crowd. To begin with she did not like her school friends attending her big events as they, uncharacteristically, tried to distract her opponents. However, her sporting success resulted in an improvement in her friends' behaviour, and she found herself relying on their presence to raise her game.

Complete the table to identify **two** types of group or social behaviour with illustrations from the paragraph.

Type of behaviour	*Illustration from paragraph*
1	
2	

2

[Turn over

DO NOT WRITE IN THIS MARGIN

Marks

12. The following data refer to concentrations of phosphate detected in water of Scottish rivers between 1986 and 2004.

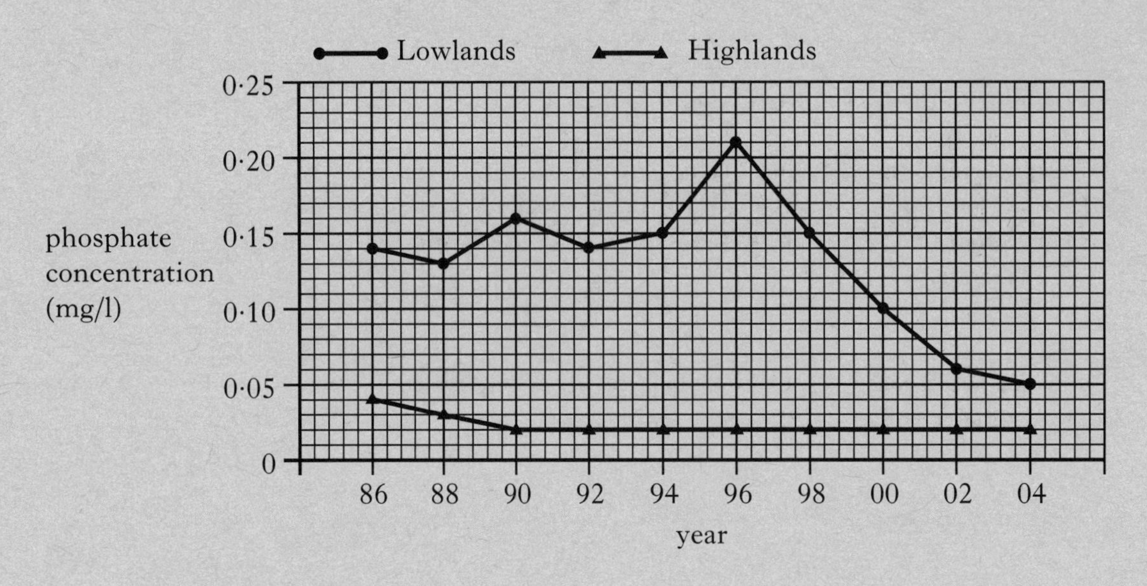

(*a*) What are the maximum and minimum concentrations of phosphate recorded in each of the areas?

	Phosphate Concentration (mg/l)	
Area	*Maximum*	*Minimum*
Highlands		
Lowlands		

1

(*b*) (i) Compare and contrast the data from 1996 to 2004.

2

(ii) Explain how the higher rainfall in the Highlands might contribute to the difference between the phosphate concentrations of rivers in the Highlands and Lowlands.

1

DO NOT WRITE IN THIS MARGIN

Marks

12. (*b*) (continued)

(iii) Suggest a reason for the change in phosphate concentration of the rivers in the Lowlands since 1996.

____

____ **1**

(*c*) (i) Aquatic plants absorb phosphate from river water against the concentration gradient. What term is used to describe this process?

________________________ **1**

(ii) What type of molecule in the plant cell membrane is involved in this process?

________________________ **1**

[Turn over

DO NOT WRITE IN THIS MARGIN

Marks

SECTION C

Both questions in this section should be attempted.

Note that each question contains a choice.

Questions 1 and 2 should be attempted on the blank pages which follow.

Supplementary sheets, if required, may be obtained from the invigilator.

Labelled diagrams may be used where appropriate.

1. Answer **either** A **or** B.

A. Describe the functions of the liver under the following headings:

(i) production of urea; **2**

(ii) metabolism of carbohydrates; **5**

(iii) breakdown of red blood cells. **3**

(10)

OR

B. Describe the cardiac cycle under the following headings:

(i) nervous and hormonal control of heart beat; **4**

(ii) the conducting system of the heart. **6**

(10)

In question 2 ONE mark is available for coherence and ONE mark is available for relevance.

2. Answer **either** A **or** B.

A. Give an account of the transmission of a nerve impulse at a synapse. **(10)**

OR

B. Give an account of the carbon cycle and its disruption by human activities. **(10)**

[*END OF QUESTION PAPER*]

DO NOT WRITE IN THIS MARGIN

SPACE FOR ANSWERS

DO NOT WRITE IN THIS MARGIN

SPACE FOR ANSWERS

DO NOT WRITE IN THIS MARGIN

SPACE FOR ANSWERS

DO NOT WRITE IN THIS MARGIN

SPACE FOR ANSWERS

ADDITIONAL GRAPH PAPER FOR QUESTION 7(*c*)(ii)

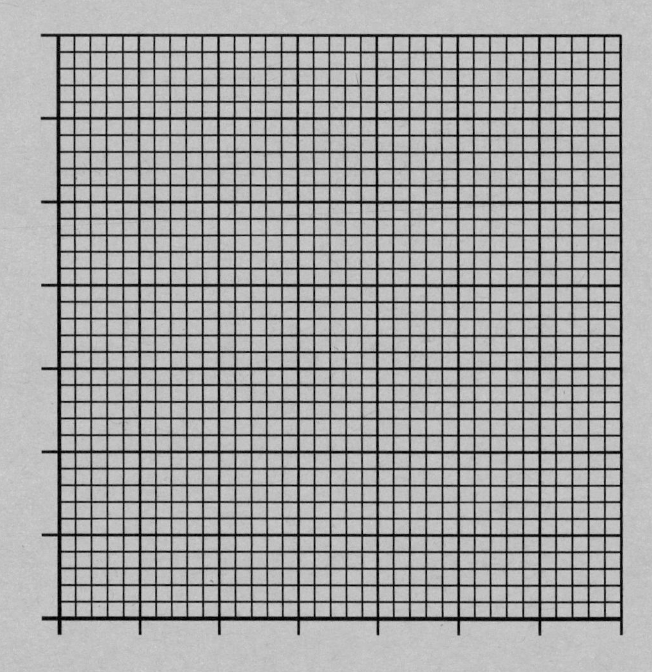

[BLANK PAGE]

[BLANK PAGE]

[BLANK PAGE]

[BLANK PAGE]

[BLANK PAGE]

Acknowledgements

Leckie & Leckie is grateful to the copyright holders, as credited, for permission to use their material:
The Scotsman for two bar graphs (2002 p 8).

The following companies have very generously given permission to reproduce their copyright material free of charge:
HarperCollins for the diagram taken from *Advanced Human Biology* by J. Simpkins and J.I. Williams (2003 p 14).